CONTRACTING DETAILS

A do-it-yourself construction schedule and homebuilding handbook

SCOTT WATSON and
SHEILA HOLLIHAN ELLIOT

BAINE BOOKS
East Troy, Wisconsin

© 1999 by Scott Watson and Sheila Hollihan Elliot. All rights reserved. No part of this book may be reproduced, stored in a retrieval system or transmitted in any form, or by any means, electronic, mechanical, photocopying, recording, or otherwise, without permission of the publisher, except by a reviewer, who may quote brief passages in a review.

Published by: Baine Books
                Post Office Box 892
                East Troy, WI 53120

Publisher's Cataloging-in-Publication Data
Watson, Scott
    Contracting details: a do-it-yourself construction schedule and homebuilding handbook / by Scott Watson and Sheila Hollihan Elliot.
        Includes index.
        ISBN 0-9667837-0-0 (pb.)
        1. House construction-Amateurs' manuals. 2. Building-Amateurs' manuals. 3. Contractors-Amateurs' manuals. I. Elliot, Sheila Hollihan. II. Title.
TH4811.S47 1999
690.837 — dc20   LCCN 98-74459

Printed in the United States of America

Cover and text design by Tamara L. Dever, TLC Graphics, Orangevale, CA
Cover illustration by Josh Glasgow

Table of Contents

PART ONE: PLANNING
Page 1

Step One	Getting started	3
Step Two	Defining your goals	9
Step Three	Creating a plan	25
Step Four	Locating qualified professionals	39
Step Five	Setting the schedule	57

PART TWO: BUILDING
Page 65

Step Six	Preparing for construction	67
Step Seven	Preparing the site	73
Step Eight	Building a solid foundation	78
Step Nine	Constructing the shell	92
Step Ten	Completing the rough work	104
Step Eleven	Closing interior walls	116
Step Twelve	Trimming it out	127
Step Thirteen	Finishing construction	137
Step Fourteen	Wrapping it up	143

PART THREE: APPENDIXES
Page 149

Appendix A	Telephone and address list	150
Appendix B	Cost breakdown sheets	152
Appendix C	Construction calendar	156
Appendix D	Utility hotline numbers	162

INDEX
Page 164

ACKNOWLEDGMENTS

The authors wish to express their thanks and appreciation to
Guy Ramstack of Ramstack Design and Drafting L.L.C.,
to Marcus Costopoulos of Marcus Built and to Josh Kent of
Kent Building Company, Inc., who reviewed portions of this book
and contributed their professional suggestions and comments.

Also, a special thanks to Peter Pappa, Judith Rasmussen,
Candy Stapleton, Anne Tuttrup and Sue Fieber for the hours
they spent reading and critiquing.

And finally, this book would not have been possible without
the patience, love, support and encouragement of friends and family.
Thank you John and Gilbert Elliot.
Thank you Adam, Ian, Nicholas and Emily Watson.

Visit the authors at their web sites.

Sheila Hollihan Elliot:
http://www.SheilaElliot.com

Scott Watson Construction:
http://www.BaineBooks.com

WARNING — DISCLAIMER

This book is designed to provide information in regard to the subject matter covered. It is sold with the understanding that the publisher and authors are not engaged in rendering legal, accounting or other professional services. If legal or other expert assistance is required, the services of a competent professional should be sought.

It is not the purpose of this handbook to reprint all the information that is otherwise available to the authors and/or publisher, but to complement, amplify and supplement other texts. You are urged to read all the available material, learn as much as possible about home construction and tailor the information to your individual needs.

Every effort has been made to make this handbook as complete and as accurate as possible. However, there may be mistakes in both typography and content. Therefore, this text should be used only as a general guide and not as the ultimate source of home construction information. Furthermore, this handbook contains information on home construction only up to the printing date.

The purpose of this handbook is to educate and entertain. The authors and Baine Books shall have neither liability nor responsibility to any person or entity with respect to any loss or damage caused, or alleged to be caused, directly or indirectly, by the information contained in this book.

If you do not wish to be bound by the above, you may return this book to the publisher for a full refund.

WELCOME!

General contracting your own home is an exciting, rewarding and creative project. As a general contractor, you will not be building the house yourself, but rather managing the building process.

Imagine taking control, building exactly the house you want and saving money. A professional builder, no matter how well intentioned, can never do as accurate an assessment of your needs as you can. And by avoiding a professional contractor's profit and overhead fees, you can save 10 to 20 percent. On a budget of $100,000 for labor and materials, that is at least a $10,000 savings!

This workbook takes you step-by-step through the entire building process. The same concepts, schedules and forms used by professionals are included here, but with detailed guidelines and complete checklists to make them suitable for the novice.

If you are willing to take the time, make careful plans and follow each step through methodically, you can successfully general contract your new home.

PART 1

• •

PLANNING

*The results of the project can be
no better than the plan.*

GETTING STARTED
DEFINING YOUR GOALS
CREATING A PLAN
LOCATING QUALIFIED PROFESSIONALS
SETTING THE SCHEDULE

Project managers will tell you that the most important part of any project is the plan. A well-developed building plan will produce a well-built house on time and within budget. Incomplete plans promise unwanted surprises, additional expenses and increased stress. Why take the chance? Spend time now getting prepared, making decisions and developing a detailed home construction plan.

In Part One, you will be guided through the planning process. First, use the worksheets to help you define your needs and locate the perfect house plans and building site. Next, spend time locating, interviewing and hiring the right professionals to finance and build your home. Finally, create a building schedule to use during the construction phase.

The result of all this hard work will be a complete construction plan that balances quality, features and price. So get out your pencil and get started.

STEP 1

GETTING STARTED

- *Understand your role as a general contractor*
- *Use this book*
- *Get organized*

Efficiently managing your new home construction project depends upon a complete understanding of the construction sequence and the duties of the general contractor.

In step one, you will discover the varied responsibilities the general contractor assumes. You will see how this book assists you in handling these responsibilities, and you will learn how to keep detailed and organized records.

UNDERSTAND YOUR ROLE AS A GENERAL CONTRACTOR

Think of a general contractor as a coach. The general contractor, or GC, assembles the best possible team of professionals and then manages that team so that everyone comes out with a win — the GC with a beautiful new home, the workers with paychecks in their pockets and everyone with the satisfaction of a job well done.

As head of the construction team, having a basic understanding of how a house is built is necessary, but attempting to acquire the skills each of the building trades uses is neither practical nor efficient and only duplicates the responsibilities of the subteams, or subcontractors, you will carefully select. Crews appreciate and respect a general contractor who efficiently manages the job but does not attempt to become a crewmember. It is probably to your advantage not to be skilled in these various tasks so that you will not be tempted to step out of your GC role.

As the project manager, you need to impart confidence and motivation by accurately scheduling the various subcontractors, arranging for timely deliveries of materials and establishing good working relationships with local building officials and lenders. You must inspect and approve quality workmanship, secure corrections without damaging team spirit and promptly pay for completed work. And finally you must have lots of energy and enthusiasm for the project.

Managing the project so that it progresses smoothly may sound easy, but be aware that general contracting takes time and energy. So ask yourself if you can be available to schedule work, coordinate details, and make daily visits to the job site to inspect progress and workmanship. Also know that even the best-planned, most efficiently run project will experience problems. Consider your ability to face challenges and make clear decisions before choosing to self-contract. If you have concerns, you may want to consider hiring a professional builder to consult with at various stages in the building process.

Just remember to keep things in perspective. Anyone choosing to build a home must plan the entire project anyway. A general contractor does not usually step in until house plans are selected, land is purchased, specifications are created and financing is secured. This book enables you to blend these homeowner duties with the construction phase and general contracting responsibilities to make building your new home an enjoyable and rewarding experience.

> Don't forget to keep a photo journal and scrapbook of your homebuilding project.

USE THIS BOOK

First, read this book thoroughly. Although you could open to page one and get started, it makes more sense to understand the entire general contracting process before you begin. The last thing you want is for your construction project to feel like a mystery novel with a surprise ending.

The home construction process must occur in a logical order called a construction sequence. Within the construction sequence are a number of steps with one step needing completion before the next step can begin. For that reason, this book is divided into steps rather than chapters. The first page of each step will list the various duties to be performed before moving on to the next step. In some cases these duties work best if they are performed simultaneously, while others must be done one at a time and in order. In both instances, the directions will be clear.

The construction sequence as presented here will be accurate for most locations, but variations can occur due to weather conditions, geographic landscape, trade union or common local building practices, regional building codes or regulations, and community habits. Some common variations include survey requirements, lender payment schedules and building department inspection requirements. You will be reminded to take notice of these possible variations throughout the planning phase, and will be prompted to write them down immediately in the proper areas of this book.

Note that at various stages throughout the book there are worksheets and forms for you to complete. These help you to plan and manage the project, and give you a permanent record of all your decisions. In addition, each step contains a detailed checklist and areas to make notes of dates, conversations and other important information. Every successful general contractor uses tools — not hammers and nails, but paper and pencils. These schedules, forms and worksheets complete your toolbox.

Note that this book is not intended to be an encyclopedia of building techniques, skills and practices, but rather a concise construction schedule. However, it is important to have a basic understanding of structural elements. If you think you need more background information, go to the public library and ask the librarian to point you toward books on building and construction. Study diagrams of the parts of a house and see if you can identify the purpose of each. Think about what would happen to the total structure if any part were missing or poorly constructed. Study photographs and drawings that will help you recognize quality in your new home construction.

Using this schedule and clearly documenting all variations helps you avoid confusion, and results in a well-managed project.

GET ORGANIZED

Organization makes the job run more smoothly. Besides using this book and writing everything down, you should also purchase a large three-ring binder and several divider sheets. Make a category for each of the things listed on the next page. Do not worry if you do not yet understand the meaning of each category. As you move through the steps you will discover what each thing is and be reminded to include it in your organizational binder.

CATEGORIES FOR AN ORGANIZED THREE-RING BINDER

- Financial papers
- Forms to submit for draws
- Building department approved plan
- Working copy plan
- Spec sheets including kitchen cabinet layout
- Contracts and proposals
- Paperwork returned with building permits
- Copy of percolation test
- Change order forms
- Lien waivers
- Receipts
- Invoices
- Correspondence from anyone involved with project
- Surveys, both proposed and existing
- Cost breakdown
- Phone and address list

Accurate record keeping is an important responsibility for every general contractor, so keep your records in order and with you at all times. That way you will be able to handle questions and make decisions as necessary throughout the day.

Now that you understand the homebuilding process and the roles and responsibilities of a general contractor, you are ready to proceed to the next step.

STEP ONE CHECKLIST

____ 1. I understand that my duties as a general contractor include:

- scheduling subcontractors
- arranging for materials delivery
- managing cash flow
- getting building department permissions and inspections
- approving workmanship
- handling problems and conflicts

____ 2. I understand that general contracting my home will take a great deal of time, attention and energy, and I am prepared to make daily visits to the job site to inspect progress.

____ 3. I have a basic understanding of the structural elements of a house.

____ 4. I have read this entire book and understand the general contracting process.

____ 5. I understand that home construction must follow a logical building sequence but that variations do occur and I should make notes and alterations as necessary in this book.

____ 6. I understand the importance of keeping accurate records and will complete the worksheets, checklists and notes sections throughout the project.

____ 7. I have purchased a three-ring binder and have an organizational plan.

NOTES

STEP 2
DEFINING YOUR GOALS

- *Plan for your needs*
- *Pre-qualify for financing*
- *Estimate building costs*

It is easy to get caught up in the excitement when you are planning a building project. Sprawling model homes and flashy sales brochures lure you to make high-priced purchases.

Step two helps you resist the urge to over-build and over-spend by having you define your needs and your budget before you start shopping for house plans. You will visit this step many times throughout the planning stage to help you remain focused. The tasks for this step can be done in any order or simultaneously.

PLAN FOR YOUR NEEDS

Before you begin to plan your new home, you must first consider your lifestyle, clarify your needs and define your wants. You probably have lots of ideas of what would be nice to have in a new home, but often these ideas are vague and are not prioritized. The worksheets included in this step will help you organize your thoughts.

LIFESTYLE WORKSHEET

It is easy to take your life and all you have for granted or to forget about certain things that are important to the way you live. The Lifestyle Worksheet helps you define aspects of your life that may or should influence the kind of home you build. When answering these questions, think not only of the present, but anticipate future needs and changes.

How many adults will live in the home? Consider the possibility of grown children moving back home or an elderly relative sharing your home.

_____

_____

How many children will live in the home? Be sure to consider the ages of the children and how the needs for space may change as the children grow in number and age.

_____

_____

How many and what kinds of pets do you own? A pet used to running free on a large piece of property may not adjust well to city living. Also consider changes you may need to make in your lifestyle to accommodate dog-walking schedules and so forth.

_____

_____

How long do you plan to live in this home? If you plan to live in this home for only a short amount of time, you may want to consider the resale aspects of your new home more than personal satisfaction with the layout and style.

_____

_____

How many and what kind of vehicles do you own? It would be easy to forget about the snowmobile while planning

---

Remember that property taxes vary from location to location. Consider the impact this will have on your personal finances.

during the summer months. Also, consider additional vehicles that you may buy in the future that will need storage space.

_____

_____

Do you have any special physical needs? Think about the need for wheelchair ramps or air purification systems for example.

_____

_____

Do you have a home business? Consider the additional work and storage space you may need to conduct business from your home.

_____

_____

What are the leisure activities for the adults? Consider the extra space you would need for hobbies like gardening or woodworking, and the quiet areas necessary for reading and listening to music.

_____

_____

What are the leisure activities for the children? These activities will change with growth, so consider sports, video games, toys, bicycles, loud music and socializing.

_____

_____

What family activities do you enjoy? Entertaining, swimming and barbecuing are examples.

_____

_____

Do you have unusual time restrictions? If you spend long hours at work or if you travel frequently, you may want to include elements in your new home that require less maintenance.

_____

_____

Do you have privacy concerns? Active neighborhoods with many children will increase foot traffic and noise levels. Also, the locations of busy roads, bike paths and businesses are things to consider.

_____

_____

Other? List here anything pertinent to your lifestyle. Safety concerns, frequent weekend guests or grandchildren who spend a week each summer might all be examples.

_____

_____

_____

You can see after completing the first worksheet that considering your lifestyle is an important first step in planning for your housing needs. It would be silly for someone who loves boating to plan a home that does not have easy access to water or a place to store a boat during the off season. But by keeping the focus on your lifestyle needs, you can successfully plan the right home for yourself.

NEW HOME FEATURES WORKSHEET

Using the information from the Lifestyle Worksheet, you will begin to plan the physical aspects of your new home, realistically balancing your needs with your desires. The New Home Features Worksheet is meant to prompt your imagination as you make your plans. Make notes to yourself concerning things you definitely want to include in your new home and things you would consider upgrading or expanding if you find you have additional money in your budget and can afford some extras. Also, write down anything you need or want that has not been included on this worksheet, and cross out anything that does not appeal to you.

List your needs and preferences below each feature.

Style Consider whether you want a one-story ranch, a two-story colonial, a three-story Victorian, a split-level or some other style. Think about basement and attic space.

_____

_____

_____

Bedrooms Decide how many bedrooms you need, the size you require and the location within the house. Some people like to have the master bedroom away from the other bedrooms. Also, think about the closet space necessary in each room.

_____

_____

_____

Bathrooms Consider the number of half baths and full baths you need and their locations. List all the fixtures you will include in each and the need for cabinets and closets.

_____

_____

_____

Living space Some people are happy with one living area while others appreciate a formal living room to accompany a family room. Consider the location for any living areas within the house and the size that would work best for your needs.

_____
_____
_____

Dining areas Think about formal dining rooms, dinettes, snack bars, breakfast nooks, and their sizes and locations.

_____
_____
_____

Kitchen Cabinet space, counter space and appliances all need to be considered when planning a kitchen. Some kitchens feature work islands, desks and sitting areas, which require more space. Think about the location of the kitchen in relation to any dining areas and entrances that will be used for carrying in groceries.

_____
_____
_____

Den/Office Home computers, hobbies and home businesses often require an extra room. Consider its size and location within the house.

_____
_____
_____

Recreation room Pool tables, card tables, dance floors and hobbies may require large and strategically located rooms.

_____
_____
_____

Laundry room Laundry rooms can be located in the basement, in the mud room or near bedrooms.

_____
_____
_____

Mud room Frequently the laundry room and additional closets are incorporated into the mud room.

_____
_____
_____

Additional storage Consider the need and location for extra closet space for coats, linens, sports equipment and collectibles.

_____
_____
_____

Garage Consider the size of the garage and whether it will be attached or detached. Think about additional storage space for lawn equipment, power tools, lawn furniture and recreational vehicles.

_____
_____
_____

Patio/Deck/Porch Think about where these should be located and how big they should be.

_____
_____
_____

Other List anything you definitely want to include in your new home such as a swimming pool, a large open foyer, an exposed basement and so forth.

_____
_____
_____

You should now have a good idea of the kind of home you want to build and the features you want included. You will appreciate the preliminary work you did here when the time comes to make selections from the variety of home designs available.

LAND FEATURES WORKSHEET

When setting criteria for purchasing land, you must consider not only the physical aspects of the piece of property but also the quality of the community that the property is a part of. To complete the Land Features Worksheet, write down everything that is important to you when considering the property you want to live on. Consider the distance you would travel for work, school, recreation and emergencies. Also, think about community services, neighborhood cohesiveness and property values.

List your needs and preferences below each feature.

Location Consider whether you would rather live in a rural, an urban or a suburban area.

_____
_____
_____

Size Think about how much time you can commit to a piece of property for maintenance and care. Also consider the activities you may enjoy that require more outdoor area.

_____
_____
_____

Privacy considerations Think about neighborhood demographics, area businesses and the nearness of buildings on adjoining lots.

_____
_____
_____

Landscape, vegetation and view Consider whether you prefer a small square lot, a sloped lot overlooking a city skyline, a large wooded area or some other type of land.

_____
_____
_____

Community services Consider community services you need or want such as clubs, organizations, hospitals, public libraries or garbage pick-up.

_____
_____
_____

Employment Think about the distance you would consider driving to work. Or perhaps you are looking for an area that offers new employment opportunities.

_____
_____
_____

Schools Consider the distance you would consider living from schools. Think about the quality of schools you would demand and special services you require.

_____
_____
_____

Recreation Think about your need to be near parks, bike paths, ski hills, swimming pools, bowling alleys and dance halls.

_____
_____
_____

Neighborhoods Consider whether you want to be a part of a neighborhood with an active association or a private community where the neighbors barely know each other.

_____
_____
_____

Other Write down anything else that is important to you concerning property. Perhaps you want a lot that allows you to bird watch, has low property taxes or has easy access to public transportation.

_____
_____
_____

Understanding the type of property you want and the quality of the community you need will help you in future steps as you search for land to purchase.

You have just completed a very important step in your new home project — you have defined your needs and clarified your desires. You will use these clear guidelines throughout the planning process. Share the results of these worksheets with lenders, architects, realtors, subcontractors and other professionals involved in your project to assist them in better understanding your personal needs and preferences for your new home.

PRE-QUALIFY FOR FINANCING

To pre-qualify for financing means to find out approximately how much money you can borrow for a home mortgage based on your personal finances. Given your income, assets, debts and liabilities, a lender uses various formulas to determine the amount you can comfortably afford to borrow. A thorough disclosure on your part will elicit a more accurate quote, so be honest and prepared. Complete the Pre-qualify for Financing Worksheet and have it with you when you call or visit lenders to pre-qualify for financing.

PRE-QUALIFY FOR FINANCING WORKSHEET

Income (either monthly or yearly):

Employment ....$ _____

Child support/maintenance ....$ _____

Other...$ _____

Assets:

Savings..$ _____

Real estate ....$ _____

Auto ....$ _____

Household goods .... $ _____

Collectibles .... $ _____

Life insurance policies .... $ _____

Retirement funds .... $ _____

Other .... $ _____

Debts and liabilities:

Credit cards .... $ _____

Auto loans .... $ _____

Mortgage .... $ _____

Other loans .... $ _____

Child support/maintenance .... $ _____

Other .... $ _____

To pre-qualify for a loan is not the same as being pre-approved for a loan. Because the lender has not checked your financial records or the validity of your project, the lender is not guaranteeing you a loan. The lender is simply stating that if all the information you provided is complete, it is likely that you would be approved for a certain amount of money. Neither you nor the lender is obligated to commit to a mortgage at this point, so contact as many lenders as you wish. Banks, credit unions, mortgage bankers, and savings and loans are all in the business of home loans and most will offer to pre-qualify you for free — sometimes over the phone.

Using advertisements, personal preferences, the yellow pages and recommendations from friends and family, make a list of lenders you will approach to pre-qualify for financing using the Pre-qualifying Lenders Worksheet. Save yourself time and energy by collecting information about permanent financing at the same time that you pre-qualify for a loan. Terms, interest rates and policies vary from lender to lender, so gather materials now to use when making decisions later. Use the list of questions provided. Write down the responses you receive and use this information in the future to evaluate and select the lender that will finance your new home.

PRE-QUALIFYING LENDERS WORKSHEET

Name _____

Address _____

Phone _____

Notes _____

Name _____

Address _____

Phone _____

Notes _____

Name _____

Address _____

Phone _____

Notes _____

QUESTIONS TO ASK LENDERS

1. Do you have any restrictions or policies regarding a homeowner acting as general contractor? If so, what are they?

Each lending institution will make its own rules regarding who is qualified to general contract a home. Some may require that the job be completely handled by an experienced general contractor. Others may want an experienced builder to handle only the paperwork but will allow the homeowner to run all other aspects of the job. Other lenders may have no problem with a homeowner acting as the general contractor.

2. Do you offer construction loans that can be rolled into a mortgage when construction is completed?

A short-term construction loan funds the project during the building phase. As a certain percentage of the home is completed, the bank will release funds to cover that same percentage. This is called a 'draw' on the loan. There can be anywhere from four to eight draws depending on local building practices.

Once the building is completed, that short-term construction loan needs to be rolled into a regular mortgage on your new home. With a permanent mortgage, you pay interest plus the principal. In a construction loan, you pay only the interest on the amount you have drawn. The principal of the construction loan is paid with the mortgage. You can save time and money if the same lender handles both the construction loan and the permanent mortgage.

3. What mortgage packages do you offer?

RATES: Interest rates are constantly fluctuating and depend on the terms of the loan. Bargain for the best interest rate possible.

TERMS: A variety of options are available for types, lengths and conditions of loans. Bargain for the best terms possible.

DOWN PAYMENTS: Usually a certain percentage of the total project cost is expected as a down payment. If the down payment is low, you are usually required to carry mortgage insurance. Typically, property, closing costs, survey fees, house plan fees and building permit fees will qualify toward a down payment.

4. What are the average closing costs? Are these included in the mortgage, or do I need to pay cash?

Lenders will charge for appraisals, credit reports, title searches and other services that must be paid when the mortgage is finalized or closed.

5. How are payments to subcontractors and suppliers handled in a construction loan?

It is advantageous to the GC if the lender pays subcontractors and suppliers directly during a draw, rather than the GC being responsible for writing checks. Each subcontractor will need to produce a lien waiver before the money is released. This is a document stating that the subcontractor has been paid in full and has no claims against the property; however subcontractors do not want to waive the lien until they have been paid in full. This sticky situation can be avoided when the lender handles the payments upon presentation of a lien waiver.

6. *What can I do to prepare to secure financing, and what items will I need?*

   Typically, these items will include a loan application, various financial documents and all information concerning land and construction.

7. *How long will it take to secure a loan once I start the process?*

   The bank will need to gather information such as credit reports and appraisals before agreeing to finance your home; these processes take time.

Pre-qualifying for financing helps you to budget your project and prevents you from spending money on land or house plans you cannot afford. Now that you know how much money you realistically have to spend, you can estimate the amount of house you can afford to build.

ESTIMATE BUILDING COSTS

Before you can purchase land, select house plans or choose the kinds of materials you want in your house, you must have a realistic idea of what you can afford. Pre-qualifying for financing indicates the amount of money you have to work with, but it does not give any indication of the amount of house that money will build.

Consider the couple who approached a home designer with a list of all the elements they wanted included in their new home, but gave no indication of their budget. The designer developed beautiful plans, and they were happy to pay over $1,000 for them. After obtaining bids from area contractors, they discovered they were $125,000 over budget! Again they approached the designer, but this time with a list of things they wanted eliminated from the original plans. They paid for the second set of plans, obtained bids and discovered they were still $80,000 over budget. They asked the builders to explain how the prices could remain so high when they had eliminated so much, and they discovered that they had retained all the most expensive parts of the plan while eliminating relatively inexpensive items.

This couple failed to estimate building costs prior to planning and ended up wasting time and making costly mistakes. Of course exact expenses cannot be determined until final decisions are made, but there are ways of getting ballpark figures that can help you to estimate building costs in your area. Use the following suggestions to complete the Cost Estimation Worksheet.

1. Visit home shows and open houses and ask for prices, specifications and any other information that is available. Remember that the person you talk to will probably be a salesperson so ask specific questions and make sure you get detailed answers.

2. Contact builders and tradespeople and ask for general prices for their services.

For instance, call a plumber and ask for an approximate price to plumb a full bath (toilet, vanity, bathtub/shower unit) using mid-grade fixtures.

3. Visit various suppliers of building materials to get an idea of what certain items cost.

4. Some builders may give you a per square foot price that they use for estimating costs. Ask for a figure that includes high quality foundation, framing and mechanical work, but mid-grade personal choices like carpeting and kitchen cabinets.

5. Ask your local building department for a cost valuation sheet for your region. This allows you to calculate a square footage price for each part of a home (basement, bathroom, garage) based on average local fees.

6. Apply logic when considering which home features will be most cost effective. A two-story home will be somewhat less expensive than a ranch home with the same number of square feet, because less foundation and roof labor and materials are needed. Eliminating a small bedroom will not save as much money as eliminating a bathroom that needs water lines, waste lines, ventilation and fixtures. A complicated and steep roof system will cost more, as it requires additional time and materials to complete. The price goes up with every window you purchase and so on.

7. Call real estate agents for approximate land prices in your area.

8. Look at house plan books for the price of stock plans or contact home designers for their prices.

9. Contact lenders regarding financing fees and construction loan interest rates.

10. Call utility companies, building departments and others as necessary to estimate the costs for specific services.

11. In addition, ten to fifteen percent of the entire budget should be reserved in a contingency fund. This money will pay for any complications, necessary changes or additional materials not planned for.

COST ESTIMATION WORKSHEET

Contingency fund ....$ _____

Land...$ _____

House plans ....$ _____

Financing fees..$ _____

Construction loan interest...$ _____

Building permits..$ _____

Utility connections..$ _____

Well or city water connection ....$ _____

Septic or city sewer connection ....$ _____

Percolation test .... $ _____

Landscaping .... $ _____

Driveways, sidewalks, patios and decks .... $ _____

Garage .... $ _____

House square foot price .... $ _____

Excavation .... $ _____

Foundation .... $ _____

Framing .... $ _____

Lumber .... $ _____

Windows and doors .... $ _____

Roof .... $ _____

Exterior finishes .... $ _____

Drywall .... $ _____

Interior wall finishes .... $ _____

Flooring .... $ _____

Trim .... $ _____

Plumbing .... $ _____

Electrical .... $ _____

Light fixtures .... $ _____

Heating, ventilation, air conditioning (HVAC) .... $ _____

Insulation .... $ _____

Other .... $ _____

Compare your cost estimation figures to those figures you received from lenders regarding the amount of money you qualify to borrow. Remember that many factors go into determining the final cost of a home: the location, size and physical characteristics of the property; the size, shape and complexity of the house plans; and, the size, type and grade of materials used. The purpose of estimating is to ensure that you allow enough money for necessities like well, septic and foundation, while deciding if there is even a chance you can afford the hot tub. Armed with this loose estimate, you can confidently proceed to the next step.

NOTE: For this book, cost estimation is meant as just that, an estimation of labor and materials needed to complete the project. Some use the same term to identify amounts quoted by subcontractors and suppliers in written proposals, but realistically those are not estimates, but rather firm prices. Make sure when you talk about cost estimation with lenders, subcontractors and suppliers that you define what you mean by the term.

STEP TWO CHECKLIST

_____ 1. I have completed the *Lifestyle Worksheet*, the *New Home Features Worksheet* and the *Land Features Worksheet* and will use these to develop a suitable building plan.

_____ 2. I have completed the *Pre-qualifying for Financing Worksheet* accurately and honestly, and I know that lenders will use this information to indicate approximately how much money I can comfortably afford to borrow.

_____ 3. I have completed the *Pre-qualifying Lenders Worksheet* and have contacted several lenders to pre-qualify for a mortgage. At the same time, I gathered pertinent financing information to use later when selecting a lending institution.

_____ 4. I have completed the *Cost Estimation Worksheet* and now have a better understanding of various building costs in my area.

_____ 5. I understand that a cost estimation is only an estimation and that actual prices will vary according to the final building products and materials I choose.

_____ 6. I understand that the reason for cost estimating is to allow enough money in the final budget for quality structural elements before choosing less important or decorative elements.

_____ 7. I understand the importance of a contingency fund to pay for unforeseen expenses.

NOTES

STEP 3
CREATING A PLAN

- *Find land*
- *Select house plans*
- *Create specifications*
- *Contact building inspector*

At last you get to plan your new home. All decisions finalized here will make the rest of the project run more smoothly, so spend plenty of time choosing and selecting each element you will include in your home.

The four tasks in this step must be performed simultaneously, as developments in one area may cause changes or restrictions in another. For example, the slope, shape or size of a plot of land will not accommodate every style of house. Also, each house plan is unique and requires individual specifications. And it is always a good practice to visit the area building inspector to ensure that the house plans and building site you choose will complement local building practices and adhere to codes and regulations.

FIND LAND

Remember the adage, the three most important things in real estate are location, location, and location? It turns out that that is only partially true. When choosing land to purchase, you must balance a desirable location with personal preferences and building suitability. In fact, if a choice needs to be made, it is probably wiser to spend a little more on an ideal piece of property and a little less on a house. You can always upgrade a house, but with land you must ultimately buy what you want and want what you buy.

One couple met their personal and location preferences when they purchased a heavily wooded lot within easy driving distance from their work and leisure pursuits. However, they failed to evaluate the land for building suitability and were frustrated to find that the shape and slope of the lot would make all aspects of construction difficult and expensive. The shape of the lot would require extensive excavation. There was only one good entrance to the building site and that would need a complicated driveway to reach it. The builders shook their heads when they saw the potential complications this property posed. Avoid these same problems by carefully evaluating your property before purchasing.

LOCATING PROPERTY

Before searching for land, review the Land Features Worksheet you completed in the previous step. Keep in mind the overall quality of your dream location. It does not make sense to purchase numerous acres at the right price if you have already decided that maintaining a small yard is more than you care to do. Nor does it make sense to look at a small city lot when a large summer garden is your passion.

Once you are clear on the type of property you are interested in seeing, you can start your search. Here are some ways to locate available land.

Real estate agent The real estate agent you choose should be fully licensed and in good standing with the National Board of Realtors.

Newspapers Classified sections of most newspapers have areas devoted to land for sale. These listings will include land for sale by owners, information realtors will not have.

Friends, relatives and business associates Sometimes land is available but not advertised, so ask around.

Town hall and tax department Sometimes land is in default for taxes. You may be able to work out a deal with the owner.

> A percolation test is a procedure that measures the rate at which soil absorbs water and is used to determine the type of septic system necessary for property not serviced by city sewer.

Bulletin boards Check supermarkets, churches and organization halls.

Law and insurance offices Clients may be in distress or for other reasons thinking about selling land. Your inquiry may prompt interest.

Weekend drives Drive around looking for signs indicating land for sale.

EVALUATING PROPERTY

Contact local building authorities, consultants and neighbors to assist you in accurately judging a piece of property you are interested in buying. Evaluate it for technical and building suitability following these guidelines.

House plans Check to see that the house design you have selected will fit the size, shape and slope of the lot. For instance, an exposed basement would work better on a sloped lot. A narrow city lot requires a narrow house plan, and so on. Hire a professional builder or surveyor to help if you are unsure.

Covenants Many neighborhoods and subdivisions develop policies and standards called covenants. For land use, that generally means restrictions for backyard projects like outbuildings, play equipment, clotheslines and landscaping. Check to see if the neighborhood has covenants and if so that they are acceptable to you.

Septic Check to see if there is city sewer or if you will need to install a septic system. If you need a septic system, hire a contractor to perform a percolation test. The results of this test will dictate the type of system you need. File a copy of these results in your organizational binder.

Well Check to see if there is city water or if you will need to drill a well. Contact a local well contractor or the building department for more information.

Zoning Contact the local zoning department to ensure that your property is zoned residential and that the surrounding property has zoning you find acceptable. Investigate any zoning restrictions that apply to your property.

Environmental Investigate any natural conditions that could cause problems for home construction. For instance, rocky soil could be difficult and expensive to dig through. And, if neighbors experience flooded basements every year, you can assume you will too.

When you find your ideal building site you can make an offer to purchase, but your offer should have any contingencies necessary to protect you. For instance, you will not want to be responsible to purchase land if you are unable to secure financing, so the offer would be contingent upon the closing of your loan. Also, the land should have a clear title. This means that no other person has rights to or liens against your property. A title search will be a part of your financing requirements. And, because you do not know the exact soil conditions of the land, you will probably want a contingency stating that if the property is not suitable for a standard foundation, the seller will need to pay for any additional expenses. It is wise to consult with an attorney knowledgeable in real estate before purchasing property.

Spend as much time as you need to find the property that is right for you and your new home.

SELECT HOUSE PLANS

House plans need to be specific and detailed. These are the instructions that each of the subcontractors you hire will bid from and build to, so there should be no unanswered questions in your plans.

You first need to review the Home Features Worksheet that you completed in the previous step to refresh your memory about the type of house you need and want to build. It can be difficult to keep your perspective when looking at hundreds of beautiful house plans, so review your needs chart frequently. Also, review the estimates you received from lenders regarding the amount of financing you qualify for, and review the estimates you received from building professionals regarding prices for your area. Remembering to be realistic, you can begin to search for house plans.

LOCATING HOUSE PLANS

Most of us think of house plans as a bird's-eye view of the interior of a home, showing the room layout. While this is a part of it, plans, or blueprints, also describe dimensions, elevations, materials and finishes for a particular house. Each set of blueprints has many pages, each with its own important perspective and description.

Elevation Elevations are the front, rear and side views of the exterior of the house.

Foundation The foundation plan will indicate the kind of foundation, show how the house will be supported on that foundation, and show drains, reinforcing rods and all other aspects of the foundation.

Floor plan This shows the room layout for each floor and includes doors, windows, closets, stairwells, fixtures and appliances.

Section A cross section shows the entire house from the foundation through the roof, indicating the types and sizes of construction materials to be used, including lumber, insulation, exterior finishes, interior finishes and roofing.

Miscellaneous Sometimes separate pages are included for plumbing, electrical and HVAC. Given a choice, ask for all the optional elements in the plans you purchase.

You can purchase house plans from a number of different sources.

Stock plans A variety of house plans can be found in books, magazines and catalogs. These plans are structurally sound and cost effective. Because these plans are meant to be generic, they may require some modifications to meet your needs and local building codes. Have a local designer or draftsperson evaluate the design and make changes as necessary.

Architects Architects are licensed professionals schooled in all aspects of building; they are often specialized. They design a home based on information you provide. If you plan on building a unique home or have trouble locating stock plans that fit your needs, you may require the services of an architect.

Designers Designers perform the same functions as architects but they have less schooling. They too will design homes based on information you provide. A local designer familiar with area building codes and regulations is often a cost-effective choice if you are having difficulty finding plans that please you.

You will be required to pay for your plans when you receive them, but this fee may count toward your down payment. You will need twenty or more copies of your final blueprints for the bidding process in Step Four. House plans are copyrighted. Check with your architect or designer to determine the proper way to obtain multiple copies of your plan. Stock plans can be purchased in reproducible form and you may make copies as necessary for your use. Copies can be made at various copy and office shops, but prices vary drastically, so shop around.

EVALUATING HOUSE PLANS

Once you have chosen house plans, you need to evaluate them to ensure they fit into your building plan. Review the worksheets you completed in the previous step to determine if the plans accommodate your needs and your budget. If so, evaluate them for building suitability.

Lot Make sure the size, style and shape of the house plans you like will comfortably fit on the lot you plan to purchase. Check with the building department regarding setbacks. These restrict you from building too close to the lot lines.

Neighborhood Consider how well your house style and house value fit in with surrounding homes.

Covenants Make sure your house plans accommodate area standards and restrictions called covenants. Some common house covenant requirements include natural exterior finishes, a minimum square footage, a certain garage size, fireplaces and paved driveways.

Climate Different style houses are common in different geographic areas for a reason. Build the house most suitable for your climate.

Never do structural changes during construction. Remember, anytime you make any changes to plans, you must modify the master document. Make final decisions now, before you begin construction, so that all the subcontractors and building inspectors know exactly the house you are building. Take time now to consult, evaluate and plan so that your design is exactly what you want.

CREATE SPECIFICATIONS

Think of specifications as the ultimate shopping list for your new home. Specifications, or specs, are the detailed lists of structural, functional and decorative materials used to complete your new home. Creating specs takes a great deal of time and energy as you will need to visit supply stores, consult with professionals and pore through catalogs before making final decisions.

Many specifications need to be decided early in the planning process and actually become a part of the blueprints, while other specifications can be chosen later and are listed separately on a spec sheet. Specifications necessary for complete blueprints include anything relevant to the structure or style of the house, including lumber size, foundation building materials, exterior finishes, interior wall boards, windows and doors, and the types, sizes and shapes of electrical and plumbing fixtures. Specifications listed separately on a spec sheet would include functional and decorative aspects such as the style, brand, quality and color details for homebuilding products.

The more specific you can be now, the fewer problems you will encounter later in the project. For instance, deciding on the electrical fixture you want above the dining table may not seem important now, but did you know that the electrical opening necessary for a light fixture is different from the opening left for a ceiling fan, and that both are different from a combination light fixture and ceiling fan? The electrician will need to know which fixture you plan to use in order to produce an accurate bid.

Planning thoroughly now will reduce any urges to make changes later. Changes during construction often mean building alterations that cause budget problems. Just remember that you can save money by using less expensive decorative elements, but never skimp on structural elements. You can always upgrade floor coverings or bathroom fixtures in the future, but the life span of your house depends on a solid construction. Better construction equals greater value and lower maintenance needs.

Do not worry if this sounds confusing. Specification Worksheets are included to help prompt your decisions. For some decisions, it is just a matter of checking which option you prefer. For others you must write in brand, style, size and color. Fill in the worksheets completely and use them when consulting with house designers and subcontractors. In this way, all the professionals involved with your project will know exactly what is being planned and built. Labor and material bids will be consistent, and you can avoid cost overruns and confusion. Place final spec sheets in your organizational binder.

SPECIFICATION WORKSHEETS

Foundation

Type:	☐ full basement ☐ slab	☐ exposed basement	☐ crawlspace
Material:	☐ poured concrete	☐ block	☐ pre-cast concrete
Damp proofing:	☐ rubber	☐ asphalt	☐ reinforced asphalt

Floor system

Framing:	☐ dimensional lumber	☐ truss joist	☐ engineered trusses
Sheathing:	☐ 3/4" tongue and groove plywood ☐ 3/4" tongue and groove OSB		

Exterior walls

Framing:	☐ 2"x4"	☐ 2"x6"	
Sheathing:	☐ 1" foam	☐ 1/2" OSB	☐ other_____
Options:	☐ house wrap		

Roof

Framing:	☐ cut roof	☐ engineered trusses
Sheathing:	☐ 1/2" or 5/8" OSB	☐ 1/2" or 5/8" plywood

Shingles:	☐ cedar	☐ slate	☐ tile
	☐ rubber	☐ asphalt 3-tab	☐ asphalt dimensional
	☐ other_____		
Vents:	☐ ridge vents	☐ mushroom vents	
Options:	☐ ice and water shield	☐ metal valleys	☐ attic storage

Exterior doors

Type:	☐ wood	☐ metal	☐ fiberglass
Style:	☐ six-panel	☐ four-panel	☐ flush
Options:	☐ side lights	☐ dead bolts	☐ kick plates
	☐ windows		

Windows and patio doors

brand _____ color _____

Type:	☐ wood	☐ clad	☐ vinyl
Glass:	☐ insulating	☐ low-E	
	☐ high performance		
Style:	☐ double-hung	☐ casement	☐ awning
	☐ sliders	☐ other	
Options:	☐ grills in glass	☐ wood grills	☐ other

Overhead doors

Type:	☐ wood	☐ metal	
Style:	☐ panel	☐ flush windows	
Options:	☐ opener	☐ keyless entry	☐ insulated
	☐ windows		

Exterior finishes

Soffit and fascia:	☐ wood	☐ aluminum	☐ vinyl
	style_____		color_____
Siding:	☐ wood	☐ aluminum	☐ vinyl
	☐ other_____		
	style_____		color_____
Trim:	☐ wood	☐ aluminum	
	☐ other_____		
	style_____		color_____
Stucco:	☐ conventional	☐ other_____	
	style_____		color_____
Masonry:	☐ brick	☐ stone	
	style_____		color_____

Electrical

Service type: .... ☐ overhead	☐ underground

Service amps: .... ☐ 100	☐ 200	☐ other

Needs: .... ☐ dryer	☐ range	☐ phones
☐ ceiling fans	☐ computer	☐ TV

List special needs: _____

_____

HVAC

Heating: .... ☐ electric	☐ hydronic	☐ forced air
☐ other_____

Fuel supply: .... ☐ natural gas	☐ LP gas	☐ oil
☐ other_____

Options: .... ☐ humidifier	☐ air conditioning
☐ air purifier	☐ electronic air cleaner

Gas piping

Needs: .... ☐ dryer	☐ range	☐ fireplace
☐ water heater	☐ furnace
☐ other_____

Plumbing

KITCHEN
Sink style: .... ☐ cast iron	☐ solid surface	☐ stainless
☐ other_____

Faucets: ...... brand _____ style _____ color_____

Options: .... ☐ dishwasher	☐ disposal	☐ icemaker

LAUNDRY ROOM
Location: .... ☐ basement	☐ first floor	☐ second floor
Options: .... ☐ laundry tub	☐ drain tray

MASTER BATH
Toilet: ...... brand _____ style _____ color_____
Sink: .... brand _____ style _____ color_____
Tub: .... brand _____ style _____ color_____
Shower: ..... brand _____ style _____ color_____
Faucets ..... brand _____ style _____ color_____

BATH ONE
Toilet: ...... brand _____ style _____ color_____
Sink: .... brand _____ style _____ color_____
Tub: .... brand _____ style _____ color_____
Shower: ..... brand _____ style _____ color_____
Faucets ..... brand _____ style _____ color_____

POWDER ROOM
Toilet: ...... brand _____ style _____ color_____

Sink: . . . . brand _____ style _____ color_____
Faucets: . . . . . . brand _____ style _____ color_____

BASEMENT BATH
Timeframe: . . . . ☐ future   ☐ now

HOT WATER HEATER
Type: . . . . ☐ gas   ☐ electric   ☐ direct vent gas

OTHER
Hose bibbs: . . . quantity _____
Water softener: . brand _____ style _____
Sump pump: . . brand _____ style _____

Fireplace

Type: . . . . ☐ masonry   ☐ prefab

Options: . . . . ☐ gas log   ☐ gas lighter   ☐ blowers

Finish: . . . . ☐ raised hearth   ☐ flush hearth   ☐ mantel
                       ☐ brick   ☐ tile   ☐ marble
                       ☐ other_____

Drywall

Finish: . . . . ☐ smooth   ☐ sand   ☐ plaster
                       ☐ texture   ☐ other_____

Options: . . . . ☐ primer

Cabinets

Type: . . . . ☐ custom   ☐ prefab

Species: . . . . ☐ oak   ☐ maple   ☐ hickory
                       ☐ cherry   ☐ pine
                       ☐ other_____

Door style: . . . . ☐ flush   ☐ raised panel   ☐ glass
                       ☐ arch   ☐ cathedral

Options: . . . . ☐ sliding shelves ☐ door hardware
                       ☐ crown mold   ☐ plate rail
                       ☐ cutting board   ☐ other_____

Countertops

Type: . . . . ☐ laminated   ☐ solid surface   ☐ tile
                       ☐ other_____

Trim

DOORS
Species: . . . . ☐ pine   ☐ oak   ☐ poplar
                       ☐ other_____

Style: . . . . ☐ six-panel   ☐ four-panel   ☐ flush

STEP 3

DOORKNOBS
Type: ....☐ conventional ☐ lever
    brand _____ style _____ color_____

CLOSET SHELF AND POLE
☐ standard .... ☐ other_____

BASE, CASING, BASE SHOE
Species: ....☐ pine ☐ oak ☐ poplar
                    ☐ other_____

Style: ....☐ ranch ☐ colonial ☐ fluted
                    ☐ other_____

Size: .... base _____ casing _____

Options: ....☐ crown ☐ chair rail
                    ☐ other_____

SPINDLES, POSTS AND RAILS
Species: ....☐ oak ☐ fir
                    ☐ other_____
                    style _____ size _____

OTHER
Towel bars: .... style _____ size _____
Paper holder: .... style _____ size _____
Door stops: .... style _____ size _____

Paint and stain colors

Exterior _____

Interior staining _____

Interior painting _____

Flooring (brand, style and color)

	CARPET	VINYL	TILE	WOOD	OTHER
Living					
Family					
Dining					
Dinette					
Kitchen					
Den/office					
Stairs					

	CARPET	VINYL	TILE	WOOD	OTHER
Hallways					
Foyer					
Laundry					
Master bedroom					

Bedroom one

Bedroom two

Bedroom three

Master bath

Bathroom one

Powder room

Other

Appliances (brand, style and color)

Built-in oven

Range

Cook top

Dishwasher

Refrigerator

Microwave

Dryer

Washer

Other

Miscellaneous (notes/details)

Security

Central vacuum

Shower door

Mirrors

Sidewalks

Patios/decks

Driveway

Landscaping

Well

Septic

Sound system

CONTACT BUILDING INSPECTOR

Your local building department holds the key to what you need to know about local building codes and regulations. You will be required to obtain a building permit before construction can begin. This, along with frequent inspections by the building department, is the ultimate verification that your building project is safe and does not infringe on the rights of neighbors. Call the local building inspector and make an appointment. Take your house plans, specifications and property information to the appointment and verify that your building plans are reasonable and viable before you make costly and final decisions. And just as you did with possible lenders, use this first contact as a time to collect information for future use.

QUESTIONS TO ASK THE BUILDING INSPECTOR

1. *Do I need to be licensed to act as my own general contractor?*
   Most areas will not require you to be licensed to general contract your own home.

2. *Are there specific zoning requirements I must adhere to?*
   Setbacks are common and require you to build a certain distance from all lot lines. You may also need to install and maintain sidewalks or set your foundation at a certain height.

3. *What is the building permit fee?*
   Most often the fee is charged according to the square footage of the house including the basement and garage.

4. *What other permits will I need?*
   Besides a building permit, you may also need permits for electrical, plumbing, HVAC, well, soil erosion, zoning, sanitary and curb cutting.

5. *What do I need to submit for a building permit?*
   Besides an application, you generally need two to three copies of your house plans, two to three surveys of your lot showing the proposed house location, and heat loss calculations.

6. *How long does the approval process take?*
   Times vary depending on local procedures.

7. *What is the inspection schedule?*
   The building department requires that a certain number of inspections be completed at various stages of the building process. These may vary somewhat from location to location, but usually include an inspection following footings installation, electric service installation, foundation floor set-up, rough mechanical installation, rough carpentry, insulation installation, finish mechanical installation and one last inspection before occupancy.

8. *Does the building inspector have any advice for my project?*
   Take advantage of any information this individual can offer concerning codes, regulations, builders and property.

Remember that at this point you are not yet applying for the permits, but rather verifying that your plans are sound and preparing for the application process. When you feel that you have a building lot, house plans and specifications that are realistic and suitable for your budget, you can continue to the next step.

STEP THREE CHECKLIST

_____ 1. I have chosen property, house plans and specifications that accommodate my personal preferences and budget.

_____ 2. I have chosen property, house plans and specifications that are suitable and complement each other.

_____ 3. I have a basic understanding of blueprints and have enough copies to start the bidding process in Step Four.

_____ 4. I have protected my project by soliciting the advice of an attorney and making my land purchase offer contingent on a clear title, secured financing and acceptable soil conditions.

_____ 5. I have completed the Specification Worksheets and have developed detailed and final specifications so as not to invite changes and problems during construction.

_____ 6. I have met with a building inspector to review my building plans and received a favorable response. At the same time, I gathered important information to use later when applying for building permits.

_____ 7. I now have a complete building plan.

NOTES

STEP 4

LOCATING QUALIFIED PROFESSIONALS

- *Begin the financing process*
- *Solicit bids*
- *Evaluate proposals*
- *Hire subcontractors*
- *Hire suppliers*
- *Complete the financing process*
- *Purchase land*
- *Order survey*

Now you begin finalizing the game plan and selecting team players. Each of the tasks in this step is critical to a successful project but the careful planning you have done up to this point helps make these tasks easier.

This is the most important step before the building phase because the better the subcontractors you hire, the better the construction. The tasks in this step must be done in order. Do not rush through this step, but take your time and do your homework. It will pay off in the end.

BEGIN THE FINANCING PROCESS

Securing financing takes time. After supplying the lender with your loan application, financial statements and total building plan, you must wait while the lender checks to ensure that the project is viable. By starting the loan process early, you allow the lender to run appraisals, credit reports and titles searches without slowing you down. When you are ready with final subcontractor contracts, the lender will be ready with the financing paperwork.

Use the information you collected about lenders when you pre-qualified for a loan to evaluate these lenders. Compare the answers they gave, question anything you do not fully understand and select the lender that offers the best mortgage package. Remember you are also shopping for a lender who will help your general contracting project run smoothly and efficiently.

After choosing one or more lenders you would consider doing business with, gather all the items they indicated they require to apply for financing. While these may vary somewhat, you at least need blueprints, spec sheets, a loan application, financial statements and a sound building plan. Once you have collected everything you need to officially apply for financing, you may approach the lenders. Lenders earn interest on the money they lend, so they need good projects to lend against. Take that attitude and sell your project, showing them your detailed plan that includes a contingency reserve.

When a lender offers to do business with you, it is time to talk about the details for a mortgage and construction package. Bargain the best deal you can and be sure you understand everything. Choose the lender that will give you the best service and rates.

Do not worry if the lender offers you more money than you want to borrow. Remember that you do not have to spend the entire construction loan — you will only pay for what you build. However, having a little extra money from the beginning is easier than trying to borrow additional money mid-project, so you may want to consider this.

When the lender agrees to lend you the construction money, it will be given to you only in proportion to what you have successfully built. The lender will allow you to draw a percentage of the total only when one of the lender's representatives can inspect the construction site to see that, in fact, that percentage of the house is completed. This protects the bank. You will use the same reasoning when you pay subcontractors — pay in partials only for the portion of the contract the subcontractors have actually completed. This protects you and your project.

These partial payments are called draws and they occur at certain stages during construction. Subcontractors and suppliers

> Make sure your drywall subcontractor includes in the bid a one-year callback to repair nail pops and cracks.

submit invoices for the work they have completed. You verify that these invoices are accurate and submit a request for a draw from the lender. The lender sends a representative to inspect the site to also verify that the invoices are accurate. When the work is approved, the money is released.

Four to eight draws are typical based on the lender's practice. This book is based on a five-draw schedule. The first draw follows the completion of the foundation and pays for all the work up to that point. The second draw follows completion of framing carpentry. Draw three pays for roofing, exterior finishes, rough mechanicals, water, sewer and masonry. Draw four pays for drywall and insulation. The fifth and final draw will be paid when an occupancy permit is given and will cover any remaining charges. If your lender uses more or less than five draws, adjust those dates accordingly in this book.

Your loan cannot be finalized until you have exact quotes from the contractors and suppliers you will hire. Use the time now to find the people who will construct your home.

SOLICIT BIDS

Soliciting bids is the process you use to find quality subcontractors who can give you the best service at the best price. Because homebuilding requires a large variety of specialized trades to complete the project, you will contact a number of people and companies in your search. Each of these companies will use the blueprints and spec sheets you provide to write a proposal that will outline labor, materials and prices.

The number of subcontractors you need and the specific duties they perform will depend upon your project. For example, an excavator will install culverts, but not every building site will require a culvert. And some foundation contractors will do their own excavation and site preparation, eliminating one subcontractor from your list.

The Subcontractor List included here outlines the various subcontractors and the services they typically perform as well as their bidding practices. Choose the subcontractors and services you need to complete your project. Note that not all the trades bid in the same way. Some trades bid on just the labor, and the general contractor must provide the materials. Other trades will include the materials in their labor prices. In such cases, the subcontractors are usually making a profit on the materials. If they did not, the price for labor would increase, so it is not worth arguing this point. Occasionally, a subcontractor who normally bids just labor will provide materials as a service to the customer. If you choose this option, you can ask to have the bill itemized to see which part of the total cost is for materials. These variations are described for each subcontractor.

Use your Specification Worksheets and the Subcontractor List to determine which subcontractors you will need for your project and what should be included in each bid request.

SUBCONTRACTOR LIST

Excavator This bid will include both labor and materials.

An excavator provides the following services: tree stump and brush removal; culvert installation; construction driveway installation; trench digging; foundation site preparation; backfilling the foundation; hauling dirt; rough grading the lot; finish grading the lot; and reshaping of landscape.

It is helpful, for scheduling purposes, if your foundation subcontractor also does the excavation.

Mason This bid will include both labor and materials.

A mason provides the following services: footings installation; cement block foundation installation; basement windows installation; rebar installation; drain tile installation; damp proofing foundation walls; insulating foundation walls; interior and exterior brick installation; and other masonry finishes installation.

Concrete subcontractor This bid will include both labor and materials.

The concrete subcontractor provides the following services: footings installation; poured foundation walls installation; basement windows installation; garage and basement floors installation; sidewalk, driveway and stoop installation; drain tile installation; damp proofing foundation walls; and insulating foundation walls.

Framing carpenter This bid will include labor only or labor and materials. Indicate which you prefer and ask to have the bid itemized.

The framing carpenter installs the following: interior and exterior walls; floor decks; roof rafters; roof sheathing; exterior walls sheathing; windows; exterior doors; siding; fascia; and soffit.

The framing carpenter is your key subcontractor. Try to find one who acts as GC on other jobs and consider hiring this subcontractor early for advice and recommendations during planning and hiring. This subcontractor is the logical choice to step in should some unforeseen development take your attention away from your GC role.

Roofer This bid will include labor or labor and materials. Indicate which you prefer and ask to have the bid itemized.

A roofer installs the following: roofing materials; gutters; and exterior flashing.

Siding and other exterior finishes Bidding depends upon subcontractors used.

Which subcontractor or subcontractors you need to install exterior finishes depends upon the materials you choose to cover the outside of your home. Masons will install all masonry finishes; siding subcontractors or framing carpenters will install aluminum, vinyl or wood siding; other companies may be required to install specialized finishes such as stucco.

Plumber This bid will include both labor and materials.

A plumber provides the following services: water lines installation; drain and waste lines installation; waste line vent installation; plumbing fixtures installation; washing machine, dishwasher and ice maker connections; and all gas connections.

Separate bids should be requested from the plumber for the following, as necessary for your home: hydronic heat (boiler, tubing, radiant fins and radiators); well pump and pressure tank (if the well digger does not include); and city sewer and water connections.

Ask the plumber to indicate how much of the bid is an allowance for fixtures. You will need this information if you plan to upgrade fixtures as you can afford to do so later in your project. Also ask your plumber to verify where the plumbing lines will enter, whether it be through the foundation or under the footings. Let the foundation subcontractor know this information.

Septic system This bid will include both labor and materials.

If city sewer is not available, you will need a subcontractor to install a septic system. This may be the same subcontractor who performed your percolation test when you finalized your property selection, it may be your plumber or it may be a separate septic system subcontractor. The subcontractor you choose provides the following services: tank installation; leach bed preparation; and sewer line installation.

Well This bid will include both labor and materials.

If city water is not available, you will need a subcontractor to dig a well. A well digger provides the following services: well digging; well pump installation; water line installation; and pressure tank installation. The actual price will depend upon how deep the well must be dug, so an exact quote cannot be given. Local well diggers will be able to estimate costs based on experience.

Heating, venting, air conditioning subcontractor This bid will include both labor and materials.

The HVAC subcontractor provides the following services: duct work installation for all bathroom vent fans and dryer vents; metal chimney cap installation for prefab fireplaces; sheet metal fire stopper installation; roof flashing installation; class B chimney installation; air conditioning installation; and heat loss calculations.

For forced air heating, the services also include: heat duct installation; cold air return duct installation; furnace installation; and thermostat installation.

If your area does not provide natural gas, you will need to rent a gas tank to be installed on your building site. Ask your HVAC subcontractor for specifics.

Electrician This bid will include both labor and materials, except for light fixtures.

An electrician provides the following services: electric service installation; circuit breaker box installation; wiring for all interior and exterior outlets, switches and light fixtures; light fixture installation; 220 volt installation for dryers and electric ranges; fireplace blower installation; vent fan installation; in-wall cable and telephone line installation; well pump and sump pump connections; and electric heat installation.

Insulator This bid will include both labor and materials.

Insulators provide the following services: blown-in ceiling insulation; floor insulation for slab foundations; wall insulation; and vapor barrier installation.

Drywall subcontractor This bid will include both labor and materials.

The drywall subcontractor provides the following services: drywall installation; drywall taping and finishing; primer application; plaster finish; and application of various textured finishes.

Trim carpenter This bid will include labor or labor and materials. Indicate which you want and ask to have the bid itemized.

A trim carpenter installs the following: interior trim; interior doors; stair rails; kitchen and bathroom cabinets; countertops; hardware (doorknobs, drawer pulls, towel bars, paper holder); shelving; and closet poles.

Painting and other interior wall finishes Bidding depends upon subcontractors used.

Which subcontractor or subcontractors you need to install interior finishes depends upon the materials you choose to cover your interior walls. Painters will paint interior and exterior walls and trim; a paperhanger will be needed to hang wallpaper; and other companies may be required to complete specialized finishes.

Overhead garage door This bid will include both labor and materials.

The garage door subcontractor installs the following: overhead garage door; keyless entry; and garage door opener.

Landscaper This bid includes labor or labor and materials. Indicate which you prefer.

Landscapers provide services for the following: lawn seeding; planting; paving material installation; terracing; and yard maintenance.

Develop a list of possible subcontractors for each of the trades you will be using. Ask friends, family, neighbors and coworkers for recommendations. Inquire at lumberyards, cement plants and other supply houses. Seek the opinions of others closely associated with the building trades, like local inspectors, lenders and other builders. Your framing carpenter can usually recommend other subcontractors. Take these recommendations seriously, as building a home takes a fair amount of teamwork and cooperation between subcontractors. Use the Subcontractor Worksheet to list companies recommended for your project.

SUBCONTRACTOR WORKSHEET

Company Name _____

Trade _____ Phone _____

Recommended by _____

Company Name _____

Trade _____ Phone _____

Recommended by _____

Company Name _____

Trade _____ Phone _____

Recommended by _____

Company Name _____

Trade _____ Phone _____

Recommended by _____

Company Name _____
Trade _____ Phone _____
Recommended by _____

Company Name _____
Trade _____ Phone _____
Recommended by _____

Company Name _____
Trade _____ Phone _____
Recommended by _____

Company Name _____
Trade _____ Phone _____
Recommended by _____

Company Name _____
Trade _____ Phone _____
Recommended by _____

Company Name _____
Trade _____ Phone _____
Recommended by _____

Company Name _____
Trade _____ Phone _____
Recommended by _____

Company Name _____
Trade _____ Phone _____
Recommended by _____

Company Name _____
Trade _____ Phone _____
Recommended by _____

Company Name _____
Trade _____ Phone _____
Recommended by _____

Company Name _____

Trade _____ Phone _____

Recommended by _____

Company Name _____

Trade _____ Phone _____

Recommended by _____

Company Name _____

Trade _____ Phone _____

Recommended by _____

Company Name _____

Trade _____ Phone _____

Recommended by _____

Choose two or three subcontractors from each specialized trade that you want to have bid your project and give them a call. Briefly describe your project and ask if they are interested in bidding. If the answer is yes, set up a time to meet to drop off the plans. When meeting, you will get a feel for the company and individuals you may be working with. Include a cover letter with each set of plans specifying what you want returned with the bids.

INFORMATION TO BE INCLUDED WITH BIDS

Written proposal This is a detailed description of the labor to be provided, the materials to be supplied and the total cost for both. Ask that the proposal include everything necessary to complete the project satisfactorily. Accept only written proposals. Find out how long this proposal is valid.

References Each subcontractor should provide a list of references. This list should include the names, addresses and phone numbers of customers that they have completed work for within the past six months.

Insurance information Each subcontractor needs sufficient public liability coverage and every employer needs worker's compensation insurance. The GC must carry these insurances if the subcontractors do not. You may also want to verify with the insurers that the accounts are paid and in good standing. If a policy is up for renewal before your project will be completed, include a clause in the contract specifying that it will be renewed.

Warranties A variety of warranties may be offered on labor and materials. These guarantees may come from the subcontractor, suppliers or manufacturers. Find out what warranties apply to your project. Labor should be guaranteed for quality workmanship.

Lead time Subcontractors earn money when they are able to schedule work so that they can go from job to job with little or no wait time in between. Each will need a certain amount of advance notice to schedule your project. This amount of time varies according to the trade, time of year, economy and so on. Find out how much lead time each subcontractor requires.

Permits Ask if any special permits are required for the work that that subcontractor will do. If so, find out which permits these are and who supplies and pays for these permits.

Payment Describe the payment schedule you will be using with your lender. Ask the subcontractors if they can work within those guidelines.

Policies Ask subcontractors to describe their policies, especially their change order policy. You have carefully selected house plans and created specifications so there is no reason for change. However, changes are occasionally necessary. You may choose to upgrade something close to the end of the project when you know your contingency fund is still intact. Or a subcontractor may recommend a change. In either case, investigate to make sure that the change is in the best interest of the project and your expectations for a home. Ask other subcontractors if they would recommend the same changes and find out why. Verify that the change continues to follow local building regulations. And most important, find out if this change will require other changes throughout the project. Remember that your plans are final documents. Any changes to them should be written out on a formal change order contract and signed by you and the subcontractor involved. In addition, all other subcontractors and inspectors need to be made aware of these changes.

When you drop off the cover letter, blueprints and spec sheets, ask how much time is needed to complete the bid. Usually two or three weeks are sufficient. When you are notified that the bid is complete, you may pick up the written proposal, the information you requested, and the blueprints and spec sheets. Only the subcontractors you hire need a set of plans, and you can return those later when you sign contracts.

EVALUATE PROPOSALS

As the GC it is your job to hire the best team to construct your home. The better job you do hiring, the better the project as a whole. So take your time and find the right people to make your project a success

Each subcontractor that you solicit a bid from will return to you a proposal. A proposal is an unsigned contract specifying at the very least: the names of the parties involved; change order policy; property location; payment schedule; labor, materials and prices; and warranties. By signing and returning this proposal, you have contracted with a subcontractor. Any unsigned proposal is just that — a proposal. Before you can decide whom to hire, you must evaluate each proposal.

PROPOSAL EVALUATION GUIDELINES

1. Check each proposal to ensure that everything you wanted this subcontractor to bid is included. Also, check to see if anything is included that you did not intend to have included. If that happens, call the subcontractor to find out why. It could be you omitted a necessary element.

2. Check proposal against proposal for each specialized trade. Look to see that labor and materials are consistent.

3. Check prices and throw out any bids that are drastically high or drastically low as compared to the others. Remember that you are looking for quality service at a reasonable price — not the lowest bid.

4. Check all references thoroughly. Consider first those subcontractors with the highest recommendations. Call the references provided and inquire about the quality of construction, overall satisfaction with the job, compatibility, cleanliness, scheduling and prices. Close by asking references if they would hire that subcontractor again. Also check with the building department, the Chamber of Commerce, the Better Business Bureau and trade organizations to see if any complaints have been registered.

5. Consider how compatible you felt when you met with each subcontractor. Ask yourself if you would feel comfortable working with that person or company.

6. You may want to strictly adhere to your building schedule and hire your subcontractors based on your timeline, or you may postpone your project to accommodate the schedules of key subcontractors. The choice is yours.

7. Subcontractors are entitled to operate their businesses using policies they deem acceptable. Most will be consistent but make sure they include copies of their change order forms. Also, make certain they are willing to abide by the payment schedule set up by your lender.

8. Hire only subcontractors who carry sufficient public liability and worker's compensation insurance.

9. A warranty should be realistic. It should list the materials, services, workmanship and time period covered. It should specify whether the warranty is for replacement, repair or refund. Also, find out who carries the warranty and what you must do to ensure it.

Enough cannot be said about the importance of carefully selecting your subcontractors. Careful reference checks will produce the information you need to locate qualified and conscientious subcontractors.

HIRE SUBCONTRACTORS

When you choose a subcontractor you want to hire, do not sign the contract and send it back, but rather call and set up another meeting. At this meeting, go over the proposal line by line and question everything you do not understand. Be sure to review policies regarding changes, financing, warranties and insurance policies. A contingency should be in place that the contract is void if you fail to secure adequate financing. If everything is acceptable to you, go ahead and sign the contract.

Show the construction schedule in Part Two of this book to all of the subcontractors and ask if they anticipate any variations to this schedule; be sure to write any variations down. Ask the subcontractors how much time they will need for each phase of construction that they are responsible for completing. You will need this information later when you complete a construction calendar. At the same time, tell them where and when they can reach you if they have questions.

As GC, you will also have policies, and the subcontractors need to be aware of them. Let them know that you will not be interfering with crews but that you are fully capable of inspecting and approving quality and that you will make daily site visits to inspect progress. Also, let each subcontractor know that you understand that there are factors, such as weather and labor strikes, that can delay a project, but that you expect service within a reasonable time frame. Quality subcontractors will continue to receive work from satisfied professional builders but you will probably only hire these subcontractors once. The tendency may be to put your job off while satisfying other customers, so you may need to be persistent with your expectations.

In addition, make sure all subcontractors know that it is their responsibility to maintain a safe and clean building site. Cleaner is safer. A construction site during the working hours must be clean enough to be safe or liability insurance may be invalidated. Floors should be swept, garbage and debris should be put into dumpsters, materials jutting out should be flagged, and holes and stairwells should be railed. OSHA, the Occupational Safety and Health Administration, could inspect at any time. Any violation fines go to the subcontractors and the GC. These fines can be large, so it is important to make your subcontractors contractually responsible for any fines incurred at the job site.

Complete the following organizational duties after signing contracts with subcontractors.

1. Make working copies of all the contracts and insert them into your organizational binder. File the originals in a safe place.
2. Give each subcontractor you hire a complete set of blueprints and spec sheets.
3. In the appendix of this book you will find a cost breakdown sheet. Use this form to enter all the proposed prices for each subcontractor.
4. Write down all the lead times that the subcontractors indicated they would need. Putting this information directly into this book in the proper area will help when scheduling work during the construction phase.
5. Call all subcontractors you do not hire and thank them for their time and interest.

HIRE SUPPLIERS

As you already know, many subcontractors supply their own materials; however, as the GC, you are responsible for ordering and supplying all remaining materials. The building products, supplies and materials you purchase will depend on the specifications you created and want included in your home. The number and kinds of companies you purchase these items from will depend on how varied and specialized your choices may be.

You probably visited a number of different lumberyards, showrooms and supply companies while you were creating your specifications. You must now choose the exact suppliers you will use. Ask subcontractors which supply companies they like to work with and find out why. Visit supplier showrooms and speak with sales associates at each one to get a feeling for the service that you can expect from each company. Some suppliers will offer many different building product choices, while others remain very specialized. You will use more than one company to fulfill your product needs, but try to limit yourself for convenience's sake.

The Building Products Supplier List included here outlines various suppliers and the products and services they typically provide. Use your Specification Worksheets with this list to determine which suppliers you will need to complete your project.

BUILDING PRODUCTS SUPPLIER LIST

Lumberyard Limit yourself to one lumberyard. If you split your lumber order between two or more lumberyards, you will not receive the whole house discount. Give one lumberyard a complete set of plans and spec sheets and ask for an estimate of the whole material supply. The lumberyard will complete a materials take-off, which is a list of everything you need. Because this takes several hours to complete, you will probably be charged for this service, but the money will be refunded if you hire that supplier. You can use that take-off later to solicit estimates from other lumberyards. In that way you will be receiving consistent estimates.

A lumberyard provides the following materials: framing lumber; trusses; windows; doors; siding; roofing; and trim lumber. You can also purchase the following materials from some lumberyards if you choose: flooring; countertops; cabinets; paint; wall coverings; tile; and hardware for doors, etc.

The materials lists and prices you receive from lumberyards for lumber will not be firm quotes, but rather estimates. You may require more or less lumber depending on many factors, but these estimates are done in good faith. Also, lumber is a commodity so the prices constantly fluctuate. For this reason, lumber prices will not be guaranteed. It is wise to add an additional five percent to the lumber estimates in case prices increase.

Steel beams Steel support beams may be necessary to carry the weight load of your home. If your plans call for steel beams, contact a supplier and ask to have a supplies, delivery and installation bid. Steel prices remain fairly constant from supplier to supplier so competing bids may not be necessary.

Light fixtures Purchase light fixtures from a lighting supplier. Take plenty of time to shop around for the supplier that will offer you the products you want at the best price. After choosing exactly the fixtures you want, you may purchase them and have them delivered when you need them. Your electrician will take care of installation.

Countertops and cabinets Many suppliers offer cabinets and countertops in a variety of materials, styles, qualities and brands. The company you use will draw a kitchen layout including all the special features you want to include and will also measure for the countertops. If you choose a nontraditional countertop, such as polished granite, have installation and delivery included in the bid. Include kitchen cabinet layout in your organizational binder.

Flooring The suppliers you need for flooring depend on the kinds of floor coverings you choose. Some flooring suppliers offer more than one option. Ask for these bids to include installation.

Custom stairs Carpenters can build stairways, but many people prefer custom stairs and railings. Ask your supplier to include installation in the bid. Contact your lumberyard, millwork shops or the yellow pages for company names.

Appliances Purchase appliances through local supply stores. Ask subcontractors to install and connect the appliances as necessary.

Shower doors and mirrors Purchase shower doors and mirrors through local supply stores. Ask your supplier to include installation and delivery in the bid.

Fireplace Many options are available for prefab fireplaces. Shop around to various showrooms and pick out the fireplace you want for the best price. Give that supplier a plan and ask for a proposal. The proposal should include: fireplace; chimney if one is needed; installation; and any accessories available or that you choose.

Sound systems, security systems and central vacuum systems These are popular elements in many new homes. Ask for installation to be included in the price.

When you have chosen the materials and suppliers you wish to use, ask the supply companies to provide you with the following information:

Written proposal or estimate This is detailed information about the style, size, type, color or quality of the product you may purchase, as well as the price. Ask for the builder's discount on all products you purchase.

Warranties A variety of warranties may be offered on materials and installation. Find out if a guarantee comes from the supplier or the manufacturer and what exactly the warranty covers.

Ordering Find out how much advance notice is needed for purchasing materials and supplies. Some materials will be stock items and always available, while others need to be specially ordered.

Payment Describe the payment schedule you will be using with your lender. Ask if the supplier will work within those guidelines.

Policies Ask about any policies regarding delivery charges, returned items, damaged items, restocking fees and anything else pertinent to the purchase, delivery, installation and life span of materials.

Ask how much time is needed to complete your bid. A few days are usually sufficient for most, but some suppliers may need more time. If a supplier kept a copy of your plans to use for making calculations, retrieve it at the same time you pick up your supply estimate. You can reuse the same set of plans as you go from supplier to supplier.

When you are ready to hire a supplier, contact that supplier and set up an account, so you do not have to pay cash for each purchase. At the same time, indicate who can charge to that account. You will want your framing carpenter to be able to pick up items from the lumberyard as necessary, for instance.

Complete the following organizational duties after receiving proposals and estimates from suppliers.

1. Make working copies of each proposal and insert them into your organizational binder. File the originals in a safe place.

2. Enter the cost figures for each supplier on the cost breakdown sheet located in the appendix of this book.

3. Write down all the lead times that the suppliers indicated they would need for ordering and delivering.

COMPLETE THE FINANCING PROCESS

You are now ready to complete the financing process. Provide your lender with the firm quotes you received from subcontractors and suppliers. Ask the lender when the loan paperwork will be completed and set up a date to close the loan.

At the closing, your lender will have many documents for you to read and sign. These documents state the conditions for the loan, and you should read and understand everything contained in them before signing. You will receive copies of each document you sign. Make copies of these to put into your organizational binder and file the originals in a safe place.

PURCHASE LAND

Now that you have secured financing, you can purchase your property. Work with your attorney and lender to pay for the real estate and to transfer the title to your name.

If you have not yet done so, you will need to purchase insurance. Talk to your insurance agent and lender to determine what constitutes adequate coverage. You will at least need to protect yourself with liability insurance should someone get hurt on your property. In addition you will need a homeowner's policy that covers potential hazards such as fire, flood and theft.

ORDER SURVEY

A survey is a legal description of a lot. You must now hire a surveyor to draw a plot plan, or plat. This scale drawing shows the outline of the lot, the proposed location of your home on that lot, setbacks, easements and topography. Meet the surveyor at the building site and indicate where and how you want your home positioned. The surveyor will drive stakes into the ground where all corners of your foundation are to be located.

This survey is necessary to obtain building permits and to verify compliance with zoning laws and restrictions. This survey is also necessary for the excavator, who must know where to dig for the foundation. Either the surveyor or excavator will pound additional or "offset" stakes. These stakes are positioned three to five feet behind the foundation corner stakes and indicate where the excavator should dig. While the foundation will be built according to specification, the hole will be big enough for the foundation subcontractors to complete their work.

Now that your financing has been secured, your lot professionally surveyed and each of your team players carefully selected, you can begin setting your construction calendar.

STEP FOUR CHECKLIST

_____ 1. I have carefully compared all the information I received from lenders in a previous step and have selected the one I believe will offer me the best loan package and service.

_____ 2. I have gathered all necessary items and have applied for a home construction loan and mortgage.

_____ 3. I understand that during the construction phase, a draw system is used to pay for labor and materials. I know that the number of draws necessary and the times they occur may vary from lender to lender so I must note these variations in this book.

_____ 4. I have solicited bids from the exact subcontractors I need to complete my project. At the same time, I gathered information about their policies and procedures to use later when evaluating subcontractors to hire.

_____ 5. I have carefully evaluated each proposal that I received from subcontractors. I compared the proposals I received to my spec sheets and subcontractor list to be sure that nothing was omitted.

_____ 6. I have hired the subcontractors I will use based on ability to complete project, positive references, compatibility, price and other service factors.

_____ 7. I have hired and set up accounts with all the suppliers I need to provide building materials for my home.

_____ 8. I have placed copies of each subcontractor and supplier contract in my organizational binder.

_____ 9. I have entered the proposed contract figures onto my cost breakdown sheet.

— *continued* —

STEP FOUR CHECKLIST, CONTINUED

_____ 10. I have written down the time that each subcontractor will need to complete each phase of construction.

_____ 11. I have written down all lead times the subcontractors and suppliers indicated they would need.

_____ 12. I have completed the loan process and included copies of all documents in my organizational binder.

_____ 13. I have purchased my lot and bought insurance for my protection.

_____ 14. I have had my lot surveyed and the foundation corners are staked.

NOTES

STEP 5

SETTING THE SCHEDULE

- *Apply for building permits*
- *Apply for utilities*
- *Create a construction calendar*

Up to this point, the majority of your duties have been as a prospective homeowner, but now it is time to begin your full-time general contracting duties.

In this step, you will make the final preparations before moving on to the building phase. Do the tasks in this step in order. While you are waiting for applications to be processed, you can finalize a construction calendar.

APPLY FOR BUILDING PERMITS

A building permit is the permission your municipality gives you to build your home. Building codes are developed for the health and safety of people, and permits are a way of checking to see that these building codes are being met. In addition, the building department will inspect the construction site at various times to check that the home is being built according to regulations. If the site passes inspection, work can continue. If the site does not pass inspection, the building inspector will inform you of necessary changes. Contact the subcontractors responsible for making the corrections and call for a reinspection as soon as the corrections are completed.

When you met with the building inspector in step three, you asked for a list of all the materials you would need to apply for the building permits. Gather these items now. Necessary items may vary somewhat depending on your locale, but you will at least need an application, two or more house plans and a survey showing the proposed location of the house on the lot.

Contact the building inspector and schedule an appointment to apply for permits when you have assembled all the items. Use this appointment to verify that you have included in your package all the required elements. Also verify the information you received in the first meeting and question anything you still do not understand. The building department will thoroughly check your plans to see if they are in compliance with local codes and zoning rules. Sometimes more than one department must approve plans.

You will be contacted when the review is complete. Either your plans will be approved or you will be apprised of changes and variances you must make. These changes are not optional, but they should be minor, as you have been working with the building inspector from the beginning. You must agree to build the approved plans making all noted changes. In a sense you are signing a contract with the town to perform as agreed. Inform all subcontractors of necessary changes.

At this time verify the inspection schedule for your municipality. Necessary inspections will vary from locale to locale so be sure to write down any inspections that your area requires that differ from this book. You will be reminded to schedule the following inspections that are common requirements in most areas.

1. Before footings are poured to inspect that they will rest on load-bearing soil.

2. Before backfilling the foundation to inspect proper drain tile and damp proofing installation.

3. Before pouring concrete foundation slab to inspect that

Some subdivisions require an architectural review of your plans before building approval is given.

under-floor plumbing drains, vents, interior drain tile and vapor barriers are properly installed.

4. Following electric service installation to ensure proper grounding.

5. Following completion of framing carpentry and rough mechanical installation to verify structural soundness, quality installation and proper material use.

6. Following insulation installation to check that all cavities are insulated, vapor barriers are complete and roof is vented properly.

7. Following sewer or septic line installation to check for proper flow and proper venting.

8. Before occupancy to see that wiring works properly, smoke detectors function, toilets flush, faucets work and so on.

You must pay for the permits when you receive them. Post the building permits in a prominent location on the building site. Cover them with a waterproof shield, adhere them to a board and stake the board into the ground. Place the approved plan, working copy plan and all other permit paperwork into your organizational binder.

APPLY FOR UTILITIES

Utility companies will need time to schedule your project and run service to your new home. Contact the gas, electric, telephone and cable TV companies and ask them to send you an application so you may apply for service. If city sewer and water are available, apply for these services at this time as well.

Take note of individual procedures each company requires you to follow and write service dates down in this book.

CREATE A CONSTRUCTION CALENDAR

When you hired your subcontractors, you asked them to provide you with the length of time that they would need to complete each phase of the construction project. You will use that information now, along with the detailed construction schedule in Part Two of this book, to create a building calendar. Because you do not yet have your building permits, you cannot schedule the exact day you will start, but you can create a timetable of events. When your permits are secured and your excavator has been scheduled, you can begin adding dates to your calendar.

Many factors can cause delays in your construction schedule, so be prepared and allow a little extra time. Be realistic with your expectations. Remember that this construction calendar is only tentative and you may need to occasionally adjust your dates. Use the following guidelines to help you fill in the construction calendar in Part Three of this book.

1. Enter the estimated amount of time subcontractors indicated they would need to complete each phase of construction.

2. Allow two extra days for each subcontractor whose work is outside and could be hindered by bad weather. If the framing carpenter says that your project will

take fifteen days to complete, plan a seventeen-day building schedule. Allow extra days for the excavator, foundation subcontractor, framing carpenter, exterior finish subcontractors, landscapers and driveway subcontractors.

3. Include at least one extra day for each of the large mechanical operations, such as plumbing, HVAC and electrical.
4. Allow ample time to secure inspections. You are better off letting the house sit for a day or two than having work continue before the site has passed inspection.
5. Remember that some tasks must be completed before the next can begin, and some tasks can occur simultaneously. Plan for some overlapping schedules on your calendar.
6. Include material ordering and delivery dates on your construction calendar.
7. Include lead-time dates for scheduling subcontractors.

Do not worry if this sounds confusing. Detailed instructions and examples in Part Three will assist you in completing your construction calendar.

Now that you have a tentative calendar, you will be ready to start construction as soon as your building permits arrive.

STEP FIVE CHECKLIST

_____ 1. I have gathered all necessary items, met with the building inspector and applied for my building permits.

_____ 2. I have a basic understanding of the procedures the building department will follow when evaluating my plan. I know I must abide by department rulings and make changes as necessary.

_____ 3. I understand that certain inspections will be required at certain times throughout the construction phase and that I should make notes and alterations as necessary when they differ from the schedule provided in this book.

_____ 4. I understand that if I fail a building inspection I must secure corrections and call for a reinspection before continuing construction.

_____ 5. I have posted the building permits in a prominent site on my property.

_____ 6. I have applied for all necessary utilities and have recorded dates and information these companies provided.

_____ 7. I have used the information I received from my subcontractors to create a tentative construction schedule.

NOTES

PART 2

BUILDING

Efficient management results in the timely completion of your new home.

PREPARING FOR CONSTRUCTION	COMPLETING THE ROUGH WORK
PREPARING THE SITE	CLOSING INTERIOR WALLS
BUILDING A SOLID FOUNDATION	TRIMMING IT OUT
	FINISHING CONSTRUCTION
CONSTRUCTING THE SHELL	WRAPPING IT UP

The building phase of home construction is very different from the planning phase. In Part One, you took time to carefully consult, evaluate and plan your project. During the building phase, you put those plans to work. Complete plans help ensure that any problem-solving you do will be relatively minor.

In Part Two, you will be guided through the construction process. Managing the construction schedule in a timely manner ensures progress, as each subcontractor performs necessary duties before the next subcontractor arrives. It also means that subcontractors are not waiting for needed materials to be delivered to the job site. Remember to adjust this schedule as necessary to accommodate your specific plan and local building practices.

Carefully coordinating the various schedules and making daily visits to the job site will help you manage your project more effectively.

STEP 6
PREPARING FOR CONSTRUCTION

- *Schedule excavator to prepare building site*
- *Schedule foundation subcontractor to install foundation*
- *Schedule framing carpenter to frame house*
- *Call utilities hotline*
- *Order trusses, windows, exterior doors and lumber*
- *Order a portable toilet*

Construction can begin as soon as a schedule can be coordinated with the first subcontractors. If your foundation subcontractor also does excavation work, the schedule will be easier to coordinate. The tasks in this step should be performed in order, as one scheduled construction phase affects each of the others.

SCHEDULE EXCAVATOR TO PREPARE BUILDING SITE

As soon as building permits have been secured, contact your excavator to schedule a start date for site preparation. Make this date one to two weeks in the future. This allows utilities enough time to mark the area for underground cables, but does not allow so much time that stakes get pulled out or lost. Kids love those flags.

Meet the excavator at the building site before excavation begins. Together you can check to see that the survey stakes remain in place, that the underground cables are clearly marked, and that the entrance for the driveway is known. Also, let the excavator know where to place the topsoil when the building site is scraped. Make sure the area you designate for this will not interfere with construction, deliveries and utility installations. The utility companies will not work around piles of debris or soil, so keep their pathways clear.

This is also the right time to work with the excavator to set the height of the foundation. The excavator will dig the foundation according to the height set by municipalities and the information given to you when you received your building permits. Given the elevation for the top of the foundation and a benchmark (many municipalities use the center of the road), the excavator can figure the depth to which to dig the foundation. If the municipality does not set the height, work with the excavator to determine the proper height based on the terrain of the property and the ability to spread extra excavation material evenly around the building lot.

■ Important questions to ask the excavator when you schedule site preparation:

Do I need to confirm this start date? If so, when?

Do I need to order or have any materials delivered to the building site?

How should I mark tree stumps and shrubs that are to be removed?

Do I need to mark the driveway entrance to the property?

When will the site preparation be completed?

SCHEDULE FOUNDATION SUBCONTRACTOR TO INSTALL FOUNDATION

Now that you know when the building site will be ready, you can schedule the foundation subcontractor. Foundation work should begin as soon as possible following excavation so that the hole does not sit empty for too long. A hole in the ground poses

---

Sometimes the foundation floor and garage floor can be poured at the same time. Talk to your concrete subcontractor about this possibility.

safety concerns and risks potential collapse. Let this subcontractor know where plumbing lines will enter through the foundation so that openings can be made to accommodate these lines.

If you are using both a mason and a concrete subcontractor, you will need to schedule each phase of foundation construction separately to ensure that subcontractors are available as necessary. For example, if a mason is building, damp proofing and insulating the foundation walls, but a separate concrete subcontractor is pouring the foundation floor, you will need to schedule and confirm these dates with both subcontractors.

■ Important questions to ask the foundation subcontractor when you schedule foundation construction:

Do I need to confirm this start date? If so, when?

Do I need to order or have any materials delivered to the building site?

Do you know the location in the foundation walls for all the openings necessary for plumbing lines?

Will you call for the necessary inspections or should I?

Do you need a ramp into the hole? (If the answer is yes, be sure to let your excavator know this.)

When will the foundation walls be completed?

When will the entire foundation be completed?

SCHEDULE FRAMING CARPENTER TO FRAME HOUSE

When you know the estimated completion date for the foundation, schedule your framing carpenter. You want your framing carpenter to start right after the foundation is completed for continued progress and to cover the foundation hole.

As the GC, you will be ordering all materials not supplied by the subcontractor. In addition, you will be scheduling delivery of these materials to the job site as they are needed. You do not, however, want all the framing material delivered to the job site at one time. For instance, you do not want to have windows delivered the first day of framing and risk damage or theft. It is better if the windows are delivered when the framer is ready to install them. For this reason, you may want to ask your framing carpenter to schedule framing lumber deliveries as they are needed. If your framing carpenter is not willing to schedule deliveries, ask for confirmation of materials needed before scheduling deliveries yourself. Lumber lists are usually organized according to the framing sequence, and you can simply verify which of these items will be required at any certain time.

■ Important questions to ask the framing carpenter when you schedule framing construction:

Do I need to confirm this start date? If so, when?

Do I need to have any materials delivered to the building site? If so, please confirm for me the lumber you need and the days you need it.

Where should the framing lumber be stacked on the building site?

When will the first floor deck be completed?

When will framing carpentry be completed?

CALL UTILITIES HOTLINE

Utility cables often run underground and can easily be damaged when digging. You can avoid these frustrating and expensive problems by contacting your area utility hotline. You can find these numbers in the appendix of this book, or by calling the North American One-Call Referral Service at 1-888-258-0808.

The hotline operator will ask you a variety of questions regarding the property and its location. The hotline will want to know when the excavator is expected to start digging. That is why you scheduled your excavator first, so that you can answer this question. Next, the hotline will contact certain utilities that will respond by inspecting your site. If underground cables cross your site, they will be marked and your excavator will know not to dig within a certain distance of those cables.

Find out how much time the utilities need to inspect and mark the property. Usually it is about three full days. Also, find out if there are any utilities that you must contact yourself.

ORDER TRUSSES, WINDOWS, EXTERIOR DOORS AND LUMBER

Order framing materials from your lumberyard to be available for your carpenter's scheduled start date. If construction should be delayed for any reason, you can postpone the delivery of these materials to the job site. Materials like trusses, however, will be delivered directly to the job site from the manufacturer, so be sure to adjust this delivery date if necessary.

Remember to use your spec sheets and supplier estimates for ordering. When the materials are delivered, check off quantity and quality from your sheet.

ORDER A PORTABLE TOILET

You will want to have toilet facilities available on the construction site. Use the yellow pages to locate companies that provide this service. You will probably need to sign a contract to keep the facility for a certain amount of time. Look at your construction schedule and have the contract terminate when you estimate your plumbing fixtures will be in service.

STEP SIX CHECKLIST

_____ 1. I have contacted all necessary utilities to ensure underground cables are not disturbed during construction.

_____ 2. I have scheduled the excavator to begin site preparation following utility site inspection.

_____ 3. I have met with the excavator to confirm:
- the location of the driveway
- the foundation height
- the foundation corners are staked
- the location at which to pile the topsoil

_____ 4. I have scheduled the foundation subcontractor to begin construction following excavation.

_____ 5. I have scheduled the framing carpenter to begin construction following foundation completion.

_____ 6. I have discussed a materials delivery schedule with my framing contractor and supplier.

_____ 7. I have ordered all framing materials and the first delivery date is scheduled.

_____ 8. I have arranged for a portable toilet to be delivered to the job site before excavation begins.

_____ 9. I have written down important dates and other information.

NOTES

STEP 7
PREPARING THE SITE

- *Building site is excavated*
- *Order survey*

Construction finally begins. Your first subcontractor will bring in all the heavy equipment necessary to move the dirt and prepare the site for the subcontractors who will construct your home. The two tasks in this step must be performed in order.

BUILDING SITE IS EXCAVATED

Excavation is the preparation of the building site. The amount of site preparation needed will depend on the size, location and landscape of your property. For example, a small city lot will require less site preparation than a large wooded lot in a rural area. The excavator will do the following site preparation tasks as necessary according to the building plan and contract.

Construction drive The topsoil is scraped off the path designated for your driveway and a layer of stone is installed. This becomes the construction drive as well as the base for your permanent drive.

Culvert Culverts allow water to pass through ditches without obstruction so that flooding is less likely to occur. Municipalities decide where culverts are needed and what style is necessary.

Foundation First the foundation site plus an additional ten feet is cleared of tree stumps, brush and large rocks. Next, the topsoil is scraped off and set aside for later use. Then, the excavator will use the surveyor stakes as a guide to dig the foundation hole. This hole will be three to five feet larger than the actual foundation size to allow enough room for foundation subcontractors to work.

Trenches The excavator will dig trenches for the footings. If you are installing a slab foundation, the excavator will also dig trenches as necessary for such things as plumbing and HVAC.

Safety fence Some areas require that a safety fence encompass a foundation hole. Sometimes an excavator will install the fence and other times it is the responsibility of the GC. Make sure that this is taken care of.

Clean road It is the responsibility of the GC to make sure that the road is kept clean of all dirt and mud tracked from the building site. Ask the subcontractors to clean up after themselves, but if they do not, you must.

Generally, site preparation can be completed in a day. The excavator will return to backfill when the foundation is complete and to finish grade the yard at the end of the project. Ask the excavator now the amount of lead time necessary to schedule for these tasks and write those times down in this book.

ORDER SURVEY

Some areas will require you to re-survey the location of the house on the property to ensure that the foundation will be located in the proper place and within all zoning guidelines. This is sometimes called a re-certification and is more common in urban areas than rural.

> Carry a hammer, level and tape measure with you when doing daily site visits.

STEP SEVEN CHECKLIST

_____ 1. The construction drive is in its proper location. It was installed by scraping off the topsoil and covering with a layer of stone.

_____ 2. The culvert was installed according to municipality specifications.

_____ 3. The building site was properly cleared of all vegetation.

_____ 4. The topsoil was stripped from the foundation site and stockpiled in the area I designated earlier.

_____ 5. The foundation was dug according to dimensions set by the surveyor.

_____ 6. Footing trenches and other trenches necessary for my project have been dug.

_____ 7. There is a ramp into the foundation hole as necessary according to the foundation subcontractor.

_____ 8. A safety fence encompasses the foundation hole.

_____ 9. Dirt and mud tracked onto the road have been cleaned away.

_____ 10. I have had my lot surveyed and have filed the re-certification drawings in my organizational binder.

NOTES

STEP 8

BUILDING A SOLID FOUNDATION

- *Footings are set*
- *Call for inspection of footings*
- *Footings are poured*
- *Schedule plumber to install under-floor plumbing*
- *Foundation walls are built*
- *Order steel and schedule installation*
- *Foundation walls are damp proofed*
- *Foundation walls are insulated*
- *Exterior drain tile is installed*
- *Call for inspection of foundation*
- *Under-floor plumbing is installed*
- *Call for inspection of under-floor plumbing*
- *Foundation walls are braced and backfilled, and the lot is rough graded*
- *Foundation floor/slab is set*

- *Call for inspection of foundation floor/slab*
- *Steel support beams are set*
- *Foundation floor/slab is poured*
- *Order survey*
- *Request first draw*

The foundation is the most important part of the house because it supports the entire weight load of the house while anchoring it in position. If you are building your home on a concrete slab rather than a basement or crawlspace, you will have fewer steps and these are indicated for your convenience. Do the tasks in this step in order.

FOOTINGS ARE SET
(basement and crawlspace)

Footings are exactly that, the feet of the house. They carry the entire weight load of the house and must rest on load-bearing soil below the frost line to prevent shifting and heaving. Footings are made of concrete and sometimes reinforced with steel bars called rebar. They are generally at least twice the size of the element they support, so an eight-inch foundation wall requires a minimum sixteen-inch wide footing. Footings are usually eight to twelve inches thick.

Your excavator dug the necessary trenches for the footings and now your foundation subcontractor will do the following foundation construction tasks as necessary according to the building plan and contract.

Footing forms The footing area must be set up so that the poured concrete will be the right width and depth. Forming boards are set on the undisturbed soil and fastened together to make a mold for the concrete.

Rebar If rebar will be used to attach the foundation walls to the footings, these steel bars are stuck into the ground between the forms at this time. If rebar will be used to reinforce the footings, that rebar should be on the building site ready to install when the footings are poured.

Drain tile A short piece of drain tile is run through the footings to connect the exterior drain tile to the interior drain tile. These should be installed within the forming boards or be stored on site ready to install when the footings are poured.

You will need footings for foundation walls, fireplaces, concrete stoops, post pads for decks, post pads for steel beams, exterior masonry finishes and other outdoor structures needing stability. If the sewer line is to run under the footings, make certain that the line is run previous to the footings installation. Your foundation subcontractor can pour the footings as soon as the setup passes inspection.

CALL FOR INSPECTION OF FOOTINGS
(basement and crawlspace)

Because footings are such a critical underpinning for the entire house, they must be inspected before they are poured. Sometimes the foundation subcontractor will call for the footings inspection rather than waiting for the GC to do it. In that way the footings can be formed, inspected and poured all in the same day. If this subcontractor does not call for an inspection, you must. Work cannot continue until the footings are satisfactorily inspected.

> One reason lenders require an existing foundation survey is to ensure that the house is not being built on the wrong property.

The building inspector will check to see that:

- footings rest on solid load-bearing soil
- footings rest below the frost line
- footings are set for the proper width and depth
- rebar is on site or properly installed
- footing drain tile is on site or properly installed

FOOTINGS ARE POURED
(basement and crawlspace)

After your footings setup has passed inspection, your foundation subcontractor can pour the footings.

Concrete A concrete plant will deliver and pour concrete directly into the forms. After it is poured, the foundation subcontractor will level and smooth out the concrete. Then it will be allowed to cure, or harden, overnight. The next day the forms can be stripped off and the foundation walls can be built.

Rebar Occasionally rebar is necessary to reinforce the footings. Sometimes a building plan will call for rebar or sometimes the building inspector will require it after seeing soil conditions around the footings. In either case, your subcontractor will install it when the concrete is delivered.

Key-way notches A key-way notch is used to fasten a poured concrete foundation wall to the footings. If poured concrete walls are in your plans, your foundation subcontractor will probably form these v-shaped notches into the tops of the footings.

It generally will not take more than one day to set up, inspect and pour the footings, and a short amount of time the second day to strip the forms. Your foundation subcontractor can now build the foundation walls.

SCHEDULE PLUMBER TO INSTALL UNDER-FLOOR PLUMBING
(all foundation types)

After the footings are poured, contact your plumber to schedule a date to install the plumbing that will be located under the foundation floor. This date should follow as closely as possible the completion of the foundation walls. Meet the plumber at the job site before installation of under-floor plumbing begins. Let the plumber know where you will locate the water heater, pressure tank, high efficiency furnace and air conditioner and ask for the floor drain to be installed in that area.

NOTE: *If you are building on a concrete slab rather than a basement or crawlspace, you must also contact your HVAC subcontractor and your electrician to install any under-floor wiring or cabling.*

- Important questions to ask the plumber when you schedule the under-floor plumbing:

Do I need to confirm this start date? If so, when?

Do I need to order or have any materials delivered to the building site?

Do you know the location in the foundation walls for all the opening necessary for plumbing lines?

Will you call for necessary inspections or should I?

When will the under-floor plumbing be completed?

FOUNDATION WALLS ARE BUILT
(basement and crawlspace)

When the footings are cured, the foundation walls can be built. Foundation walls rest on the footings and raise the first floor far enough off the graded soil to prevent rot. Foundation walls can be made from concrete block or poured concrete and are sometimes reinforced with rebar.

Poured concrete Poured concrete walls are formed just as the footings were formed. Boards create a mold so that the concrete will be poured to the correct depth and width. After the forms are set, concrete is poured in, leveled out and allowed to cure before the forms are stripped.

Concrete block Courses of concrete block are laid and mortared together.

Foundation openings Basement windows, sleeves for plumbing pipes and sleeves for other utility connections are installed as the walls are being constructed. Also, beam pockets are formed in the top of the foundation walls so that steel support beams can be set into them.

Brick ledge A brick ledge to support exterior masonry finishes is installed at this time.

Anchoring system Local building codes dictate how a house is to be tied to the foundation. Anchor bolts and foundation straps are two commonly used systems. The foundation subcontractor should install these anchoring devices in the tops of the foundation walls.

The time necessary to build foundation walls depends upon size, complexity and choice of building materials used. Poured concrete walls take just a couple of days to form and pour, while concrete block walls may take longer depending on the size of the crew. The foundation subcontractor will return to install the foundation floor when the under-floor plumbing is completed. Ask the foundation subcontractor to schedule that date now.

ORDER STEEL AND SCHEDULE INSTALLATION
(basement and crawlspace)

Now that your foundation walls are completed, you can contact your steel supplier and schedule to have the steel beams measured, ordered and installed. Your plans indicate the height and weight of the beams necessary for your home, but an exact length measurement is required once the foundation walls are built. It is the responsibility of the GC to get the steel measured accurately. You may do this yourself, or you may ask your framing carpenter or the steel company to measure for you. The beams can be installed anytime before the foundation floor is poured.

- Important questions to ask the steel company when you schedule steel installation:

What day will the steel be installed?

Do I need to confirm this installation date? If so, when?

FOUNDATION WALLS ARE DAMP PROOFED
(basement and crawlspace)

Damp proofing is the process used to coat the outside of the foundation walls with a waterproof material to keep moisture from seeping through. There are three main foundation waterproofing substances: hot asphalt, reinforced hot asphalt and rubberized latex. Hot asphalt is the most common waterproofing material and is adequate for areas where ground water is of little concern. Hot asphalt can be reinforced with fiberglass to prevent cracking. This is somewhat more expensive, but is also more reliable if ground water is a moderate concern. Rubberized latex is the most expensive and can be used if water is a constant problem or if your foundation wall is against a slope. Each of these substances is sprayed on and allowed to dry overnight.

FOUNDATION WALLS ARE INSULATED
(basement and crawlspace)

Foundation walls are insulated by adhering rigid foam sheets over the damp proofing substance. Sometimes the entire foundation is insulated, sometimes just the area above the frost line is insulated and sometimes the foundation is not insulated at all. Your building plan and local building codes will specify your need for foundation insulation.

EXTERIOR DRAIN TILE IS INSTALLED
(basement and crawlspace)

Besides damp proofing the foundation walls, you must also create a draining system to carry water away from the footings, so that footings always rest on solid soil. A drain tile is a long piece of plastic tubing with holes at various locations that is placed all along the footings to collect water. The water then drains through the short pieces of drain tile that were installed in the footings. From there the water runs into drain tile on the inside of the foundation and is disposed of. First, a layer of stone is laid around the footings. This prevents dirt and silt from clogging drain tiles. Next, the drain tile is put along the top or along the sides of all the footings. This needs to be inspected before it is covered. Finally a twelve-inch layer of stone is carefully put over the exterior drain tile to prevent dirt and silt from clogging the holes.

CALL FOR INSPECTION OF FOUNDATION
(basement and crawlspace)

Because proper drainage and damp proofing are important to a solid foundation, both must be inspected before they can be covered with backfill. Sometimes the foundation subcontractor will call for the inspection rather than waiting for the GC to do it. In that way, the drain tile can be installed and covered the same day. If this subcontractor does not call for an inspection, you must. Work cannot continue until the foundation walls and drain tile are satisfactorily inspected.

The building inspector will check to see that:

- damp proofing is properly installed
- insulation is properly installed
- drain tile rests appropriately on or next to footings

UNDER-FLOOR PLUMBING IS INSTALLED
(all foundation types)

Plumbing is the entire water supply and waste disposal system of a house. Water comes into your home from a well or city water source, circulates to all the fixtures and appliances, and is disposed of through a septic system or city sewer. Some of this plumbing is located under the foundation floor. Your plumber will do the following under-floor plumbing tasks as necessary according to the building plan and contract.

Sanitary crock or stack Every house needs either a sanitary crock or a stack as a part of the sewer system. A sanitary crock is used when the waste lines exit the foundation through the foundation walls. A stack is used when the waste lines exit under the footings. The plumber sets the sanitary crock or stack.

Sump crock A sump crock is set to collect the water the drain tiles gather. Later a pump is installed to dispose of the water.

Pipes and vents The plumber will lay all the drain pipes, waste pipes and water lines that run under the foundation floor. Then each waste pipe will be vented.

Drains You met with the plumber earlier to indicate where all the mechanicals will be placed. The plumber will install a floor drain in this area.

Other plumbing Under-floor plumbing necessary for basement bathrooms and laundry facilities will be installed.

The drain and connections for pipes and lines will be set high enough so as not to be covered with cement when the foundation floor is poured. Installing under-floor plumbing should not take more than a day. Once the under-floor plumbing passes inspection, your plumber will not return until your home is ready for rough plumbing.

NOTE: *If you are building on a concrete slab rather than a basement or crawlspace, the under-floor HVAC and electrical should be installed at this time as well.*

CALL FOR INSPECTION OF UNDER-FLOOR PLUMBING
(all foundation types)

Once the foundation floor is poured, it would be difficult to correct any under-floor plumbing errors, so an inspection is necessary at this point. Sometimes plumbers will call for their own inspections rather than waiting for the GC to do it. If the plumber does not call for the inspection, you must. Work cannot continue until the under-floor plumbing is satisfactorily inspected.

The building inspector will check to see that:

- crocks are installed
- pipes do not leak
- waste pipes are vented properly
- floor drains are installed

NOTE: *If you are building on a concrete slab rather than a basement or crawlspace, the under-floor HVAC and electric should also be inspected at this time.*

FOUNDATION WALLS ARE BRACED AND BACKFILLED, AND THE LOT IS ROUGH GRADED
(basement and crawlspace)

The building site becomes safer and easier to work in when the trench around the outside of the foundation is filled. If the foundation walls are braced from the inside, the trench can be backfilled and the lot rough graded.

Bracing Not every foundation needs to be braced before it is backfilled, so check with your excavator or foundation subcontractor for recommendations. But if your foundation has long expanses of straight walls or if you have unusual soil conditions, the foundation walls may need to be braced to prevent collapse. This is done by placing boards against interior walls eight to ten feet apart. Then additional boards are placed at an angle with one end positioned on those wallboards about half way down, and the other end braced against a stake, a footing or the foundation wall on the opposite side. These braces are removed before the basement floor is poured.

Backfill Once the foundation walls are braced from the inside, the excavator will replace the soil around the exterior of the foundation walls in a process known as backfilling. Fill will be bulldozed in around the entire foundation in layers. This evenly distributes the weight of the dirt and therefore the stress on the foundation. Certain soil conditions may allow only partial backfilling at this time. If this is the case, the backfilling process can be completed anytime after the first floor deck is built, as the first floor deck provides the bracing necessary to prevent collapse. Also note that some subcontractors prefer to have the first floor deck completed before doing any backfilling.

Rough grade When backfilling is complete, the lot can be rough graded. In this process, the remainder of the soil taken from the foundation hole is spread evenly around the property. Some of the excavation material may be held back to finish the backfilling or to fill in around the foundation as the soil settles. Note that the topsoil is still in reserve to use for the final grade at the end of construction.

If you live in an area where termites are a problem, be sure to have a professional treat the soil before backfilling.

FOUNDATION FLOOR/SLAB IS SET
(all foundation types)

Once the under-floor plumbing passes inspection, the concrete subcontractor can prepare the foundation floor/slab so that it can be finished in concrete.

Grade Your concrete subcontractor will set the height for the top of the foundation floor/slab and grade the ground underneath. A basement floor should be at least four inches thick and rest on at least four inches of crushed stone. Forming boards are placed across doorways or other openings.

Drain tile Interior drain tiles are placed next to the footings and run to the sump crock.

Vapor barrier A polyethylene vapor barrier is carefully set on the floor. This protects your home from moisture and radon.

Wire mesh Some building codes require wire mesh to reinforce the foundation floor. Wire mesh is installed over the vapor barrier.

The concrete subcontractor can pour concrete on the prepared foundation floor as soon as it passes inspection.

NOTE: If you are building on a concrete slab rather than a basement or crawlspace, you will probably need to insulate the foundation floor and have that insulation inspected.

CALL FOR INSPECTION OF FOUNDATION FLOOR/SLAB
(all foundation types)

The building inspector does one last check before the floor/slab is poured. Sometimes concrete subcontractors will call for their own inspections rather than waiting for the GC to do it. In that way, the foundation floor can be set up, inspected and poured in the same day. If this subcontractor does not call for the inspection, you must. Work cannot continue until the foundation floor is satisfactorily inspected.

The building inspector will check to see that:

- drain tile rests appropriately next to footings
- drain tile runs to sump crock
- vapor barrier is properly installed
- wire mesh is properly installed

STEEL SUPPORT BEAMS ARE SET
(basement and crawlspace)

Steel support beams, or I-beams, are used to span large areas, carry floor loads and distribute the weight of the house evenly over the basement or crawlspace. These steel support beams replace traditional girders built from layers of lumber. Steel beams are typically cheaper and stronger than wood beams and are not prey to termites, rot or fire.

Before your foundation floor is poured, the steel support beams should be installed. A crane will place the beams into the beam pockets in the walls of the foundation. Steel columns are placed and attached to the beams. The beams are then leveled, shimmed and braced even with the tops of the foundation walls.

FOUNDATION FLOOR/SLAB IS POURED
(all foundation types)

Your concrete subcontractor can pour the slab foundation or, once the steel beams are in place, complete the foundation floor. Concrete is delivered and poured directly onto the foundation subfloor. The foundation subcontractor will spread the concrete out and level it. The floor will be finished so that there is a gentle slope from the foundation walls to the floor drain. Curing takes twenty-four hours. If you expect rain in that time, cover the site with plastic sheeting.

ORDER SURVEY
(all foundation types)

Previous surveys indicated what you intended to build. Now that your foundation is installed, a new survey, called an existing foundation survey, will show what has been built. A foundation defines the lot and is the necessary building element needed for a survey. The survey verifies that the foundation is in the right place. This survey will be filed with your municipality and it is usually needed for the first draw. Contact your surveyor immediately to set up this appointment, so that building progress is not jeopardized.

REQUEST FIRST DRAW
(all foundation types)

Collect all the invoices for work that has been completed and materials that have been installed up to this point. For this first draw, that should include excavation and foundation work. Verify that the invoices are accurate and submit a request for a draw to your lender. Usually the first draw requires more paperwork and you will probably need to submit an existing foundation survey at this time. Your lender will send a representative to the building site to also verify that the work is complete as documented. You will be notified when the funds are released. Contact the subcontractors and suppliers who submitted invoices and tell them they can receive payment when lien waivers are presented. Put copies of the lien waivers, invoices and receipts in your organizational binder.

STEP EIGHT CHECKLIST

_____ 1. Footings necessary for foundation walls, fireplaces, concrete stoops, exterior masonry finishes and post pads were set up, inspected and installed according to building codes and plan specifications.

_____ 2. I have scheduled under-floor plumbing installation to follow foundation wall completion.

_____ 3. The foundation walls were built according to building codes and plan specifications and include the anchoring system and a brick ledge for exterior masonry finishes.

_____ 4. Basement windows and sleeves for utility connections were installed in the proper locations and beam pockets are the proper size.

_____ 5. I have measured and ordered steel support beams and scheduled installation to occur before the foundation floor is poured.

_____ 6. Foundation walls were damp proofed and insulated, and exterior drain tile was installed. The foundation passed inspection.

_____ 7. I met with the plumber to indicate where mechanicals will be located.

_____ 8. All plumbing located under the foundation floor was installed according to building codes and plan specifications. Under-floor plumbing passed inspection.

_____ 9. Termite control measures have been taken.

_____ 10. Foundation walls were braced and backfilled and the lot rough graded using all excavation soil except topsoil.

_____ 11. The foundation floor was set, inspected and poured according to building codes and plan specifications.

_____ 12. Steel beams were set and braced.

_____ 13. I have had my foundation surveyed and have filed the existing foundation survey in my organizational binder.

_____ 14. Invoices have been collected and I have submitted a draw request to my lender.

_____ 15. I have verified that the road is clean of dirt and mud tracked from my building site.

NOTES

STEP 9

CONSTRUCTING THE SHELL

- *Schedule electrician to install electric service*
- *Framing begins*
- *Order dumpster*
- *Electric service is installed*
- *Call for inspection of electric service*
- *Contact electric and gas companies for connections*
- *Schedule custom stairs for measurement*
- *Schedule roofer to install roof*
- *Schedule exterior finishes installation*
- *Schedule fireplace installation*
- *Schedule sanitary system installaion*
- *Schedule water supply installation*
- *Schedule plumber to install rough plumbing*
- *Schedule HVAC subcontractor to install rough HVAC*

- *Schedule electrician to install rough electrical*
- *Schedule all other in-wall cabling companies to install rough cabling*
- *Locks and lockbox are installed*
- *Request second draw*

You will probably be surprised how quickly your home goes from a hole in the ground to a framed structure that you can walk through. Do the tasks in this step in order. Be aware that you will be scheduling many subcontractors to begin their duties immediately following framing completion.

SCHEDULE ELECTRICIAN TO INSTALL ELECTRIC SERVICE

Once the foundation is complete and the framing is ready to begin, you can schedule the electrician to mount your electric service. Schedule this date to follow the day the carpenter has built the first floor deck.

■ Important questions to ask the electrician when you schedule to have the electric service mounted:

Do I need to confirm this date? If so, when?

Do I need to order or have any materials delivered to the building site?

Will you call for necessary inspections or should I?

How long will it take to complete electric service mounting?

FRAMING BEGINS

Once the foundation is complete, your framing carpenter can construct the skeletal structure of the house. The exterior walls, interior walls, floor decks and roof are all framed and covered with sheathing during this stage called framing. The framing carpenter will do the following framing tasks as necessary according to the building plan and contract.

Anchoring system Local building codes dictate how a home is to be tied or anchored to the foundation. Using these codes, the framing carpenter will first attach a sill plate to the top of the foundation walls, then a rim board on top of the sill plate.

First floor deck Floor trusses or joists are the boards that span the area over the basement or crawlspace and are supported by the foundation walls and steel support beams. These boards will be blocked or bridged to add support and to prevent creaking. Also, the floor deck will be reinforced under bathtubs and around openings for such things as stairwells and fireplaces. A subfloor of plywood or OSB (oriented strand board) will be glued and nailed to the floor trusses or joists.

Walls The framing carpenters will construct the exterior walls using boards called studs and will then cover the exterior with plywood, OSB or rigid foam sheathing. Openings for windows and doors will be created and reinforced. Next, interior walls will be constructed, leaving openings for doors.

Additional stories Additional floor decks and walls will be constructed if the house is more than one story.

Roof framing Roof trusses or cut boards will be used to span the

---

When deciding where to place exterior electrical outlets, consider your needs for seasonal decorations.

area over the top of the house and form the roof shape. The framed roof will then be covered with plywood or OSB sheathing.

Stairs Basement stairs will be built at this time. If you have a second story, temporary stairs with safety railings will be constructed.

Windows and exterior doors Windows and doors will be placed, leveled and secured. Have cheap lock sets delivered with the exterior doors so that they can be installed for use during construction. After construction is finished, these will be replaced with your permanent lock sets.

The size and complexity of the house will determine the amount of time necessary for framing. Depending on your contract, this carpenter may return to side your home, trim your home, or build decks and other outdoor structures.

ORDER DUMPSTER

During the construction phase, you will need a dumpster to dispose of packaging materials, lumber waste and other debris. Use the yellow pages to find companies that provide this service. Order a twenty-yard dumpster and have it placed near the construction site and drive for easy access. Just make sure it will be out of the way of delivery trucks. If the dumpster gets full and is still needed, have it emptied and returned.

ELECTRIC SERVICE IS INSTALLED

The electric service for your home consists of a service box on the outside of your house connected to a circuit box on the inside of your house. These are the only two elements necessary before the electric company will run power to your home.

The electrician will put one or two circuits into the circuit box at this time and connect an outlet to them. Once the electric company runs power to your home, you can use this outlet. Up to now, your subcontractors have used generators for power, but they can soon connect to the electric service. Other circuits will be connected later, when electrical work is finished.

CALL FOR INSPECTION OF ELECTRIC SERVICE

The building inspector must check the electrician's work before the electric company can run power to your property. Sometimes the electrician will call for the inspection rather than waiting for the GC to do it. If this subcontractor does not call for the inspection, you must. Work cannot continue until the electric service is satisfactorily inspected.

The building inspector will check to see that:

- circuit box allows for adequate circuits
- service box is in a protected spot
- service box is properly installed and grounded

CONTACT ELECTRIC AND GAS COMPANIES FOR CONNECTIONS

Once your electric service box has passed inspection, you can contact the electric and gas companies. You applied for service in an earlier step, but you must now contact these companies to tell them that the site has passed inspection and that you are ready for power. They will tell you when they will run lines from their sources to your mounted service. Be sure that no debris or materials are in their way.

The gas company will dig a trench and bring a gas line underground to the house and up to an outside meter or through the foundation sleeve to a basement meter. The gas company will leave a short threaded and capped access pipe for your plumber to connect to. The electric company will run lines either overhead or underground and connect to your service box. The electric company will also install a meter. Sometimes the telephone and cable TV companies run their cables in the same trench as the electrical cable. Contact these companies to find out.

SCHEDULE CUSTOM STAIRS FOR MEASUREMENT

Ask your framing carpenter when the floor decks will be framed. Contact the custom stair company you hired and schedule measurement for the stairs as soon as the decks are complete so that manufacturing can begin.

■ Important question to ask the custom stair company when you schedule stair measurement:

Do I need to confirm this date? If so, when?

SCHEDULE ROOFER TO INSTALL ROOF

When the framer is about two weeks from completing framing, contact the roofer to schedule a date to start roofing. The roofer can begin as soon as roof framing is complete, so that the structure is waterproofed.

■ Important questions to ask the roofer when you schedule roof installation:

Do I need to confirm this start date? If so, when?

Do I need to order or have any materials delivered to the building site?

When will the roof be completed?

SCHEDULE EXTERIOR FINISHES INSTALLATION

Once you know when the framing will be complete, you can contact the various subcontractors you hired to install the exterior finishes. These subcontractors can start as soon as the framing carpenters have completed sheathing and installed windows and exterior doors, as long as they do not get in the roofer's way. When there are multiple subcontractors involved in installing exterior finishes, it is best if the mason goes first, then the siding subcontractors. If any electrical boxes or plumbing pipes will be positioned within exterior masonry walls, the plumber and electrician need to be contacted so they can complete that work first.

- Important questions to ask the exterior finish subcontractors when you schedule exterior finish installation:

Do I need to confirm this start date? If so, when?

Do I need to order or have any materials delivered to the building site?

When will the exterior finishes be completed?

SCHEDULE FIREPLACE INSTALLATION

As soon as you know when framing will be complete, you can schedule the fireplace installation. Contact the mason or prefab fireplace company to schedule a date immediately following framing, but be sure that the chimney installation does not interfere with roofing.

- Important questions to ask the fireplace subcontractor when you schedule fireplace installation:

Do I need to confirm this start date? If so, when?

Do I need to order or have any materials delivered to the building site?

When will the fireplace installation be completed?

SCHEDULE SANITARY SYSTEM INSTALLATION

As soon as you know when framing carpentry will be completed, you can schedule to have your sanitary system installed. This date can follow framing carpentry at any time, but the sooner the better. To schedule this date, contact the subcontractor you hired to install your septic system or to run the sewer lines from the street to your home.

- Important questions to ask this subcontractor when you schedule sanitary system installation:

Do I need to confirm this start date? If so, when?

Do I need to order or have any materials delivered to the building site?

When will sanitary system installation be completed?

SCHEDULE WATER SUPPLY INSTALLATION

As soon as you know when framing carpentry will be complete, you can schedule to have your water supply system installed. This date can follow framing carpentry at any time, but the sooner the better. To schedule this date, contact the subcontractor you hired to run city water lines from the street to your home, or your well digger.

- Important questions to ask this subcontractor when you schedule water supply installation:

Do I need to confirm this start date? If so, when?

Do I need to order or have any materials delivered to the building site?

When will water supply installation be completed?

SCHEDULE PLUMBER TO INSTALL ROUGH PLUMBING

As soon as you know when framing will be complete, you can contact your plumber. Schedule the rough plumbing installation to immediately follow framing carpentry. If you are installing a gas fireplace, you may want the plumber to start one day before the fireplace installation.

■ Important questions to ask the plumber when you schedule rough plumbing installation:

Do I need to confirm this start date? If so, when?

Do I need to order or have any materials delivered to the building site?

When will rough plumbing installation be completed?

SCHEDULE HVAC SUBCONTRACTOR TO INSTALL ROUGH HVAC

Contact the HVAC subcontractor when you know the date rough plumbing will begin. The plumber needs a head start but does not need to be finished before the HVAC rough work begins. Schedule this subcontractor to start a few days after rough plumbing begins.

■ Important questions to ask the HVAC subcontractor when you call to schedule rough HVAC work:

Do I need to confirm this start date? If so, when?

Do I need to order or have any materials delivered to the building site?

When will rough HVAC installation be completed?

SCHEDULE ELECTRICIAN TO INSTALL ROUGH ELECTRICAL

When you know the expected completion date for the rough HVAC installation, you can schedule the electrician. Have the electrician start rough electrical installation the day after the rough HVAC is completed.

Meet the electrician before rough work begins to walk through the home and point out where telephone jacks, TV jacks and any special electrical openings should be positioned.

■ Important questions to ask the electrician when you schedule rough electrical installation:

Do I need to confirm this start date? If so, when?

Do I need to order or have any materials delivered to the building site?

When will the rough electrical installation be completed?

SCHEDULE ALL OTHER IN-WALL CABLING COMPANIES TO INSTALL ROUGH CABLING

The final subcontractors or suppliers that need to run cabling inside the walls can be scheduled when you know the completion date for the rough electrical work. These might include central vacuum systems, sound systems or security systems. Schedule these for installation immediately following rough electrical installation.

- Important questions to ask these subcontractors when scheduling rough installation:

Do I need to confirm this start date? If so, when?

Do I need to order or have any materials delivered to the building site?

What is the expected completion date for installation?

LOCKS AND LOCKBOX ARE INSTALLED

When framing is completed and the windows and exterior doors have been set, ask your carpenter to install the lock sets or doorknobs that you had delivered. Also, purchase a lockbox for the house key and have your carpenter attach it to your home next to the door. Give the combination for the lockbox to subcontractors, the building inspector and your lender. This system protects your home from theft and vandalism, while ensuring that the home is accessible to workers.

REQUEST SECOND DRAW

Collect all the invoices for work that has been completed and materials that have been installed up to this point. For the second draw, that should include all framing and framing materials. Verify that the invoices are accurate and submit a request for a draw to your lender. Your lender will send a representative to the building site to also verify that the work is completed as documented. You will be notified when the funds are released. Contact the subcontractors and suppliers who submitted invoices and tell them they can receive payment when lien waivers are presented. Put copies of the lien waivers, invoices and receipts in your organizational binder.

STEP NINE CHECKLIST

_____ 1. I have scheduled the electrician to mount the electric service following the framing carpenter's completion of the first floor deck.

_____ 2. Framing lumber has been delivered and stockpiled in the area I designated earlier.

_____ 3. The first floor deck was completed and anchored to the foundation appropriately.

_____ 4. I have arranged for a dumpster to be delivered to the job site when framing carpentry begins.

_____ 5. The electric service was mounted and passed inspection.

_____ 6. I have contacted the gas and electric companies to connect my service. I have checked to ensure that the utilities have a clear path from their sources to my service box.

_____ 7. I have scheduled the custom stair company to measure for stairs following the framing carpenter's completion of the first and second floor decks.

_____ 8. I have scheduled the plumber and electrician to stub out connections located within exterior masonry finishes.

_____ 9. I have scheduled the following installations to begin following framing carpentry completion: roof; exterior finishes; fireplace; sanitary system; water supply; and rough plumbing.

_____ 10. I have scheduled the HVAC rough installation to follow rough plumbing by one or two days.

_____ 11. I have scheduled the rough electrical installation to follow completion of rough HVAC.

_____ 12. I have scheduled all other rough in-wall cabling to follow completion of rough electrical.

_____ 13. Framing is finished, and windows and doors have been set.

_____ 14. Locks and a lockbox have been installed and I have given the combination to subcontractors, lenders and building inspectors.

_____ 15. The job site and road have been kept clean.

_____ 16. The dumpster has been emptied and returned if necessary.

_____ 17. All leftover framing material has been returned.

_____ 18. Invoices have been collected and a request for a draw has been submitted to the lender.

NOTES

STEP 10

COMPLETING THE ROUGH WORK

- *Tentatively schedule drywall installation*
- *Roofing is installed*
- *Exterior finishes are installed*
- *Fireplace is installed*
- *Sanitary system is installed*
- *Water supply is installed*
- *Rough plumbing is installed*
- *Rough HVAC is installed*
- *Rough electrical is installed*
- *All other rough in-wall cabling is installed*
- *Call for inspection of rough mechanicals and framing*
- *Schedule insulation installation*
- *Schedule drywall installation*
- *Schedule concrete subcontractor to install garage floor*

- *Schedule garage door installation*
- *Schedule gutters installation*
- *Request third draw*

Do the tasks in this step in order. Be aware that many of your subcontractors will be working simultaneously on the job site immediately following framing. Do not panic. This is normal in the construction sequence.

TENTATIVELY SCHEDULE DRYWALL INSTALLATION

When your carpenter has completed framing the structure, contact the drywall subcontractor and tentatively schedule a start date three to four weeks in the future for drywall installation. This subcontractor will appreciate the advance notice even if the date is tentative.

- Important question to ask the drywall subcontractor when you tentatively schedule drywall installation:

When should I confirm the start date?

ROOFING IS INSTALLED

A roof system tends to match the house it covers; the more complex the house, the more complex the roof. Also, the materials you choose to cover the roof should not only be watertight but should complement the look of the house. Your framing carpenter built the frame necessary for the roof and now your roofing subcontractor will do the following tasks as necessary according to the building plan and contract.

Drip edge First a metal drip edge is put on the eaves of the home. This prevents water from leaking behind gutters or fascia.

Tar paper or roofing felt The entire roof is covered with a tar paper or roofing felt.

Shingles or tiles Shingles and tiles are made from a variety of materials and come in a variety of styles. Shingles and tiles are put on the roof in courses with each new course lapping over the previous course so that water sheds off the roof rather than running under the shingles.

Flashing Metal flashing is installed on roof edges, around chimneys and vents, and in roof valleys to prevent leaking.

Roof vents Mushroom vents or ridge vents are installed to balance the air flow with the soffit vents. These vents minimize condensation and allow heat to escape.

The size and complexity of the roof system will determine the time necessary to complete the roof. Other subcontractors may work at the job site during roof installation, but none should interfere with the roofing crew. Getting your structure watertight is your first priority.

EXTERIOR FINISHES ARE INSTALLED

Your home's exterior finishes have two functions. One is to protect the home and its occupants from the elements. The other is

> *Consider having an address stone installed in an exterior masonry finish.*

to provide character and individuality to the home. Each home is different. Your plan may call for only one exterior finish or a combination of finishes. Your carpenter built the necessary framework and now your exterior finishing subcontractors will do the following tasks as necessary according to the building plan and contract.

Soffit and fascia The soffit and fascia are positioned under the roof overhang and give the roof and house a finished look. The siding contractor or framing carpenter installs the soffit and fascia. These are vented to allow for air flow and to prevent dampness and rot.

Masonry finishes If your home calls for an exterior masonry finish, it should be installed before siding if possible. The brick or stone is laid in courses and mortared together. All openings for electrical or plumbing located within these masonry exterior finishes should be completed before masonry finishes are installed.

Siding If you chose wood, aluminum or vinyl siding, it will be installed over the sheathing and caulked where appropriate. The siding subcontractor installs trim blocks for all exterior electrical openings.

Other Other exterior finishes such as stucco are installed. Also, wooden decks can be built at this time.

Paint Primer, paint and stain can all be applied by your painter at any time during the exterior finishes phase.

The length of time necessary to install exterior finishes depends on the types of materials you are using as well as the number of subcontractors needed. Other subcontractors will be working at the job site during exterior finishes installation. This is fine as long as finishers do not get in the way of roofers.

FIREPLACE IS INSTALLED

Masonry fireplaces and prefab fireplaces can be installed as soon as the framing carpentry is completed. Your subcontractor will install the fireplace, build the venting system and provide proper fire stopping materials. Different fireplaces require different venting systems. Some gas models vent directly out the side, while others require a chimney on the roof. Building codes dictate fire stopping measures that your subcontractors will follow.

SANITARY SYSTEM IS INSTALLED

If you have city sewer, your subcontractor will dig trenches to install the lines from the road to your house and through the foundation. This subcontractor will call to have the lines inspected before covering.

If you need a private sewer system, your subcontractor will install all tanks, pumps, sewer lines and leach beds in accordance with the type of system you require according to your percolation test. This subcontractor will have the system inspected before covering. After installation, have the subcontractor supply you with a copy of the septic report and file it in your organizational binder.

WATER SUPPLY IS INSTALLED

If you have city water, your subcontractor will dig trenches to install the lines from the road to your house and through the foundation. This subcontractor will call to have the lines inspected before covering. Also, the city will install a water meter in your basement or on an outside wall.

If you need a well, your subcontractor will use large well-drilling equipment to dig your well. There are restrictions regarding locations of wells, so work with your well subcontractor to determine its best location. After installation, have the subcontractor supply you with a copy of the clean water test and file it in your organizational binder. You can now have your well pump and pressure tank installed.

ROUGH PLUMBING IS INSTALLED

During the rough plumbing stage, the plumber will install all the pipes, vents and lines that run inside walls or under floors. No connections are yet made to fixtures, the water supply or disposal systems. Rough plumbing installation should begin as soon as framing carpentry is complete. The plumber will do the following rough plumbing tasks as necessary according to the building plan and contract.

Water lines All water lines are installed throughout the house. This will include lines to bath, kitchen and laundry rooms, and connections for fixtures.

Waste lines All drains, waste lines and vents are installed. Any pipe that runs to the sewer or septic system needs to be vented to prevent gas buildup and to prevent suction so that water can flow.

Bathtubs and showers Because bathtubs and showers have the walls built around them, they are installed at this time. Other plumbing fixtures can be attached after the walls are completed, so they are not connected until the finish plumbing stage.

Gas All gas lines are installed. This includes lines needed for fireplace, laundry room, kitchen, furnace and hot water heater.

It could take several days for rough plumbing completion. The plumber can start work while other subcontractors are there. Because the plumbing lines are large, the plumber runs lines within the walls before the other mechanical subcontractors. The plumber will return to install fixtures and make plumbing connections during the finishing phase.

ROUGH HVAC IS INSTALLED

Heating, venting and air conditioning form the climate control and air-circulating system for your home. The ductwork for these systems is hidden within a home's walls and must be installed before the walls are covered. Once the plumber has begun installing the rough plumbing, the HVAC subcontractor will do the following rough HVAC installation tasks as necessary according to the building plan and contract.

Duct work The HVAC subcontractor runs all the in-wall ductwork necessary to carry heated and air-conditioned air to each room in the house. Also, return air ductwork is installed for a complete air-circulating system.

Vents This subcontractor installs bathroom vents, clothes dryer vents and exhaust vents for kitchen ranges.

Furnace The furnace is installed as well as the chimney necessary to vent the furnace.

Air conditioner The air conditioner can also be installed at this time but be aware that the air conditioner needs to be installed outdoors at the finished grade height. Level off an area at the proper height for air conditioner installation.

Thermostat A temporary thermostat can be installed and connected to the heating and air conditioning units during cool months.

Other sheet metal work An HVAC subcontractor can also provide any other sheet metal services required, such as metal chimney caps for prefab fireplaces, fire stopping inside of walls, and metal chimneys for gas hot water heater.

The HVAC subcontractor will need several days to complete rough HVAC installation. Other subcontractors will also be working on the job site. Just make sure the plumber started at least a day or two prior to the HVAC subcontractor. The HVAC subcontractor will return to make connections and install vent covers during the finishing stage.

ROUGH ELECTRICAL IS INSTALLED

Your electrician has already mounted your electric service. During rough electrical installation, this subcontractor will thread electrical wires through the floors, ceilings and walls and leave the ends of these wires near the circuit box in the basement. The wires will not be connected into the circuit box until the finish electrical stage. Once the rough HVAC installation is completed, your electrician will do the following electrical tasks as necessary according to the building plan and contract.

Wiring The electrician runs wires for all outlets, switches, light fixtures, doorbells and smoke detectors. In addition, wiring is run to electric ranges, clothes dryers, dishwashers and fireplace blowers.

Boxes The electrician boxes out each location for outlets, light fixtures and switches.

Fans Bathroom fans are installed and wired by the electrician.

Utilities The electrician runs the cables for cable TV and telephone. The electrician will not connect these services in the finishing stage, but does run the necessary cables during the rough stage.

Heat If electric heat is in your plans, the electrician will install that at this time.

Rough electrical installation will take a few days to complete. Remember there may be other subcontractors working on the job site. The electrician will return to install fixtures, outlets and switches, and to make connections during the finishing stage.

ALL OTHER ROUGH IN-WALL CABLING IS INSTALLED

If your plans call for a sound system, a security system, a central vacuum system or any other system that requires in-wall cabling, it should be installed following rough electrical installation and according to the building plan and contracts.

CALL FOR INSPECTION OF ROUGH MECHANICALS AND FRAMING

Once the interior walls are covered, it would be difficult to correct any in-wall wiring, so an inspection is necessary at this point. Have your framing carpenter walk through before the inspector arrives to make sure that sound structure was not compromised during rough plumbing, HVAC and electrical installation. Occasionally problems can be spotted and corrected before the inspector arrives. Work cannot continue until the rough mechanicals and framing are satisfactorily inspected.

The building inspector will check to see that:

- plumbing connections are tight
- all bearing points are blocked
- all mechanicals are properly installed
- fireproofing is properly installed
- structure has not been compromised

SCHEDULE INSULATION INSTALLATION

Once the rough mechanicals have passed inspection, you can call the insulator to schedule insulation installation. Have this subcontractor install insulation as soon as possible.

■ Important questions to ask the insulation subcontractor when you schedule insulation installation:

Do I need to confirm this start date? If so, when?

Do I need to order or have any materials delivered to the building site?

Will you call for the necessary inspection or should I?

When will insulation installation be completed?

SCHEDULE DRYWALL INSTALLATION

When you know the date the insulation will be completely installed, you can call the drywall contractor to schedule a start date. Make this date for when you are sure that the insulation will have been satisfactorily inspected. Drywall should not even be delivered to the building site until the insulation has passed inspection.

■ Important questions to ask the drywall subcontractor when you schedule drywall installation:

Do I need to confirm this start date? If so, when?

Do I need to order or have any materials delivered to the building site?

When will drywall installation be completed?

SCHEDULE CONCRETE SUBCONTRACTOR TO INSTALL GARAGE FLOOR

Once the roof and the exterior finishes are completed, you can contact your concrete contractor to install the garage floor. Schedule this date for as soon as possible.

- Important questions to ask the concrete subcontractor when you schedule garage floor installation:

Do I need to confirm this start date? If so, when?

Do I need to order or have any materials delivered to the building site?

When will garage floor installation be completed?

SCHEDULE GARAGE DOOR INSTALLATION

When you know the date the garage floor will be completed, you can schedule the subcontractor who will install your overhead garage door. Set this date for two days after the floor is poured to give the floor time to cure.

- Important questions to ask the garage door subcontractor when you schedule the garage door installation:

Do I need to confirm this start date? If so, when?

Do I need to order or have any materials delivered to the building site?

When will garage door installation be completed?

SCHEDULE GUTTERS INSTALLATION

Once the roof and exterior finishes are installed, you can contact the subcontractor who will install the gutters. Schedule the installation date for as soon as possible.

- Important questions to ask the gutters subcontractor when you schedule gutters installation:

Do I need to confirm this start date? If so, when?

Do I need to order or have any materials delivered to the building site?

When will gutters installation be completed?

REQUEST THIRD DRAW

Collect all the invoices for work that has been completed and materials that have been installed up to this point. For the third draw, that should include roofing, exterior finishes, fireplace, sewer, water and all rough mechanicals. Verify that the invoices are accurate and submit a request for a draw to your lender. If you have a well and septic system, you will need to submit a copy of the safe water test and the septic report. Your lender will send a representative to the building site to also verify that the work is complete as documented. You will be notified when the funds are released. Contact the subcontractors and suppliers who submitted invoices and tell them they can receive payment when lien waivers are presented. Put copies of the lien waivers, invoices and receipts in your organizational binder.

STEP TEN CHECKLIST

_____ 1. I have tentatively scheduled the drywall subcontractor.

_____ 2. The roof is completed according to the building plan. Necessary flashing and other waterproofing measures were used to ensure a watertight structure.

_____ 3. The soffit and fascia are installed and properly vented to allow for air circulation.

_____ 4. All electric and plumbing within exterior masonry finishes have been stubbed out prior to masonry installation.

_____ 5. All exterior finishes are installed according to the building plan. All joints fit snugly and are caulked as appropriate.

_____ 6. Trim blocks were installed for all exterior electrical openings.

_____ 7. Exterior painting and staining is completed.

_____ 8. Wooden decks were built according to the building plan.

_____ 9. The fireplace was installed using proper fire stopping materials and venting systems.

_____ 10. The sanitary system I am using is installed and ready for connection. If I have a septic system, I have filed a copy of the septic report in my organizational binder.

_____ 11. The water supply system I am using is installed and ready for connection. If I have a well, I have filed a copy of my clean water test in my organizational binder.

_____ 12. All water lines, drains, waste lines and vents were installed according to the building plan.

_____ 13. Bathtubs and shower stalls were installed.

_____ 14. Gas lines were run to the necessary areas including fireplace, laundry room and kitchen.

_____ 15. I have leveled an area at the finish grade height for the air conditioning unit.

_____ 16. The ductwork, vents and thermostat have all been installed according to the building plan.

_____ 17. The furnace and air conditioner have been set.

_____ 18. All electrical wiring is installed and all outlet, light fixture and switch boxes are in their proper locations.

_____ 19. All telephone and cable TV wires and jacks were installed in proper locations.

_____ 20. Bathroom fans were installed.

_____ 21. All other in-wall-cabling companies have completed in-wall cabling according to the building plan.

_____ 22. Rough mechanicals and framing have passed inspection.

_____ 23. I have scheduled insulation installation to follow framing and rough mechanical inspection.

_____ 24. I have scheduled drywall installation to follow insulation inspection.

_____ 25. I have scheduled the concrete subcontractor to install garage floor following roof completion.

_____ 26. I have scheduled the garage door installation to follow garage floor completion by two days.

_____ 27. I have scheduled gutters installation to follow roof and exterior finishes completion.

_____ 28. Invoices have been collected and I have submitted a draw request to my lender.

NOTES

STEP 11

CLOSING INTERIOR WALLS

- *Schedule painting and staining*
- *Schedule all flooring installation except carpeting*
- *Order cabinets, countertops and trim*
- *Garage floor is installed*
- *Insulation is installed*
- *Call for inspection of insulation*
- *Drywall is installed*
- *Schedule HVAC subcontractor to finish HVAC*
- *Schedule electrician to finish electrical*
- *Schedule in-wall cabling companies to finish*
- *Schedule custom stairs installation*
- *Schedule trim carpenter to install trim*
- *Schedule plumber to finish plumbing*
- *Garage door is installed*

- *Gutters are installed*
- *Schedule light fixtures delivery*
- *Schedule appliances delivery*
- *Request fourth draw*

In this step you will be scheduling all the interior finish work, while the walls are insulated and enclosed. Do the tasks in this step in order.

SCHEDULE PAINTING AND STAINING

When the drywall is scheduled to be completed, you can contact the painter to schedule interior walls and trim for painting and staining. Set this date for the day following drywall completion. Make sure the trim is available when the painter will need it. Also, schedule other wall-covering installations, such as wallpaper.

■ Important questions to ask when scheduling the painter to paint and stain:

Do I need to confirm this start date? If so, when?

Do I need to order or have any materials delivered to the building site?

When do you want the trim delivered?

When will painting and staining be completed?

SCHEDULE ALL FLOORING INSTALLATION EXCEPT CARPETING

Schedule installation for all flooring except carpeting. This date should follow painting completion by one day to ensure that paint is completely dry.

■ Important questions to ask the flooring suppliers when you schedule flooring installation:

Do I need to confirm this start date? If so, when?

Do I need to order or have any materials delivered to the building site?

When will flooring installation be completed?

ORDER CABINETS, COUNTERTOPS AND TRIM

Call your cabinet supplier and order the cabinets and countertops. Have them delivered the day the kitchen and bathroom flooring is installed. Because countertops need to be cut to fit perfectly, they sometimes cannot be measured until the cabinets are actually installed. If this is the case, then have the tops measured and ordered immediately after cabinet installation.

Call the lumberyard and order all your trim. Have trim delivered according to your painter's scheduled start date.

Remember to use your spec sheets and supplier estimates for ordering. When the materials are delivered, check off quantity and quality from your sheet.

> Consider having concrete floors saw cut to help control cracking.

GARAGE FLOOR IS INSTALLED

The concrete subcontractor can install the garage floor anytime after the roof is completed.

Grade Your concrete subcontractor will set the height for the top of the garage floor and grade the ground underneath. A garage floor should be at least four inches thick. Forming boards will be placed across the doorway and other openings. Concrete is delivered by the cement plant and poured directly onto the garage sub-floor.

Wire mesh Wire mesh should be installed to reinforce garage floors. Other reinforcing options are available including fiberglass reinforced concrete. Talk to your concrete subcontractor to help make the right choice.

Finish The concrete subcontractor will spread the concrete out and level it. The floor will be graded so that there is a gentle slope from the back wall to the overhead garage door.

The garage floor can be formed and poured in one day, but must cure twenty-four hours before the forms can be stripped from door openings.

INSULATION IS INSTALLED

Insulation requirements are dictated by local building codes and are meant to improve the energy efficiency of your home. Once the rough mechanicals are inspected and the roof is watertight, your insulator will do the following insulating tasks as necessary according to the building plan and contract.

Caulk Have your insulator caulk all holes drilled through exterior wall plates including those under the roof. These holes were drilled for running plumbing and electrical wires, but should be tightly sealed now.

Walls and floor decks Exterior walls are insulated with fiberglass batts. These batts come in rolls and fit snugly between studs. This insulation should fill all exterior wall and exterior floor deck cavities.

Vents To allow for proper air circulation, a foam material that creates an air channel is installed from inside the roof eave to inside the exterior wall. This is called proper venting.

Blown-in insulation Insulation can also be loose and blown into ceilings. This insulation can be made of fiberglass or cellulose. This insulation cannot be installed until the other insulation has passed inspection and the ceiling drywall is installed.

Vapor barrier A vapor barrier is installed over the fiberglass batts on the inside walls and ceilings to prevent moisture from collecting next to interior walls where the air is warmer.

Insulation takes a day or two to install and longer if blown-in insulation is used. Work can continue after insulation installation passes inspection.

CALL FOR INSPECTION OF INSULATION

Because insulation is critical to an energy efficient house, the insulation must be inspected before it is covered with drywall. Work cannot continue until the insulation is satisfactorily inspected.

The building inspector will check to see that:

- all areas are insulated
- vapor barrier is properly installed
- venting is properly installed

DRYWALL IS INSTALLED

Drywall is a gypsum board covered in paper and is the most commonly used interior wall covering. Once the insulation has passed inspection, you can have your drywall delivered The drywall subcontractor will do the following tasks as necessary according to the building plan and contract.

Drywall A crew of drywall hangers installs the drywall, using nails, screws or adhesive to secure the sheets to the walls and ceilings.

Joints A crew of drywall finishers will come in next and tape all the joints. Joints and nail heads are coated with joint compound, then sanded to a smooth finish when the compound has dried. This taping and sanding process will be done several times until the walls are smooth.

Plaster If you are having plaster, the plaster crew will come in after drywall installation. The plaster crew tapes the joints and covers the walls with plaster.

Finish If you are having a specialty finish applied to your walls, that will be done at this time. Also, the drywall can be primed.

The drywall process takes several days to complete, because the joint compound needs to dry thoroughly between coats. If the drywall is installed during cold weather, the drywall crew may turn the heat up to speed the process. Do not change the setting on the thermostat if that is the case.

SCHEDULE HVAC SUBCONTRACTOR TO FINISH HVAC

SCHEDULE ELECTRICIAN TO FINISH ELECTRICAL

SCHEDULE IN-WALL CABLING COMPANIES TO FINISH

SCHEDULE CUSTOM STAIRS INSTALLATION

Once you know the day the painting and staining will be completed, you can schedule the HVAC subcontractor, the electrician and all other in-wall cabling companies to finish their work. You can also contact the custom stair company and schedule stair installation. Make the start date for each of these subcontractors and suppliers at least one day after painting is completed to ensure that the paint will be dry.

■ Important questions to ask when scheduling finish work:

Do I need to confirm this date? If so, when?

Do I need to order or have any materials delivered to the building site?

When will finish work be completed?

SCHEDULE TRIM CARPENTER TO INSTALL TRIM

Once you know the day the painting and staining will be completed and the flooring will be installed, you can schedule the trim carpenter to install trim. Make this day at least one day after painting is completed to ensure that the paint will be dry.

■ Important questions to ask the trim carpenter when you schedule trim installation:

Do I need to confirm this date? If so, when?

Do I need to order or have any materials delivered to the building site?

Will you call to have the countertops measured or should I?

When will trim work be completed?

SCHEDULE PLUMBER TO FINISH PLUMBING

Once you know when all flooring except carpeting will be installed, you can contact the plumber to schedule finish plumbing. Make this date any time after flooring is installed, but the sooner the better.

■ Important questions to ask the plumber when you schedule finish plumbing:

Do I need to confirm this date? If so, when?

Do I need to order or have any materials delivered to the building site?

When will finish work be completed?

GARAGE DOOR IS INSTALLED

Once the garage floor has had a day or two to cure, the overhead garage door can be installed. This subcontractor will install the door tracks, the door and the tension spring. This subcontractor will then install the garage door openers and keyless entry. Make sure the remotes and instructions for changing codes are left with you.

GUTTERS ARE INSTALLED

Gutters can be installed as soon as the roof and exterior finishes are completed. The gutters subcontractor will attach the gutters to the fascia and then attach all the downspouts to carry the water away from the foundation.

SCHEDULE LIGHT FIXTURES DELIVERY

You selected light fixtures during the planning stage. Have those light fixtures delivered to the job site so that they are there when the electrician arrives to finish the electrical.

Remember to use your spec sheets and supplier estimates for ordering. When the materials are delivered, check off quantity and quality from your sheet.

SCHEDULE APPLIANCES DELIVERY

You selected appliances during the planning stage. Have those appliances delivered to the job site so the plumber and electrician can make connections when finishing.

Remember to use your spec sheets and supplier estimates for ordering. When the materials are delivered, check off quantity and quality from your sheet.

REQUEST FOURTH DRAW

Collect all the invoices for work that has been completed and materials that have been installed up to this point. For the fourth draw, that should include insulation, drywall, gutters and garage completion. Verify that the invoices are accurate and submit a request for a draw to your lender. Your lender will send a representative to the building site to also verify that the work is complete as documented. You will be notified when the funds are released. Contact the subcontractors and suppliers who submitted invoices and tell them they can receive payment when lien waivers are presented. Put copies of the lien waivers, invoices and receipts in your organizational binder.

STEP ELEVEN CHECKLIST

1. I have scheduled interior painting and staining to follow drywall completion.

2. I have ordered all flooring except carpeting to be installed following painting and staining completion.

3. I have ordered cabinets, countertops and trim to arrive according to the trim carpenter's and painter's schedules.

4. The garage floor is installed to the proper height. It contains the necessary reinforcing and is finished to the appropriate slope for drainage.

5. The insulation is installed. Exterior holes were caulked and proper venting was created.

6. The vapor barrier was properly installed.

7. Insulation installation passed inspection.

8. Drywall was installed and the joints and nail heads finished properly with a smooth finish.

9. Plaster or texture finishes were installed according to specifications.

10. Drywall was primed and ready to receive paint or other interior wall coverings.

11. I have scheduled the HVAC subcontractor, electrician, in-wall cabling companies and custom stairs supplier to finish work following painting completion by one day.

12. I have scheduled the trim carpenter to install trim following painting, staining and floor installation.

13. I have scheduled the plumber to finish plumbing following floor installation.

— continued —

STEP ELEVEN CHECKLIST, CONTINUED

_____ 14. The garage door is installed according to the building plan. The remote and instructions are in my possession.

_____ 15. Gutters were installed according to the building plan.

_____ 16. I have scheduled the light fixtures to be delivered according to the electrician's schedule.

_____ 17. I have scheduled the appliances to be delivered according to the electrician's and plumber's schedules.

_____ 18. Invoices have been collected and a request for a draw submitted to the lender.

NOTES

STEP 12
TRIMMING IT OUT

- *Painting and staining are finished*
- *Flooring except carpeting is installed*
- *HVAC is finished*
- *Electrical is finished*
- *Custom stairs are finished*
- *In-wall cabling is finished*
- *Trim is installed*
- *Schedule shower doors and mirrors installation*
- *Plumbing is finished*
- *Schedule finish grade*
- *Schedule driveway, sidewalks and patios installation*
- *Schedule landscaping*
- *Schedule carpeting installation*
- *Create a punch list and schedule subcontractors*
- *Carpeting is installed*

Mechanical connections are made and your house is beginning to look like a home. The tasks in this step need to be completed in order.

PAINTING AND STAINING ARE FINISHED

The painter can begin work as soon as drywall installation is completed. After covering windows and doorways, the painter will spray the walls. Next, the trim will be painted or stained. After the trim has been installed, the painter will return to fill nail holes and touch up the paint. The number of rooms and amount of trim specified in your plans will determine the length of time necessary to complete this process. Other wall coverings, such as wallpaper, can be installed at this time as well.

FLOORING EXCEPT CARPETING IS INSTALLED

Once paint has time to dry all flooring, except for carpeting, can be installed.

Flooring suppliers will install flooring according to the building plan and contracts. If vinyl or tile floors are in you plans, an underlayment will be installed over the subfloor to provide a solid base. Hardwood is installed over a paper barrier and may need to be finished later in the construction process. Ask the supplier for more details. The length of time necessary to complete flooring installation depends upon the kinds and amounts of flooring to be installed.

HVAC IS FINISHED

Once the interior walls are finished and flooring is installed, the HVAC subcontractor will return to finish HVAC. Heat, air conditioning and return air registers are installed. The HVAC subcontractor may have installed a temporary thermostat during rough installation. If that is the case, a permanent thermostat is installed at this time. If you chose a digital thermostat, make sure you let this subcontractor know the temperature settings you will use, so that they can be programmed for you. Your HVAC system is now complete and functional.

ELECTRICAL IS FINISHED

Once the flooring is installed and the interior walls are finished, the electrician can return to finish the electrical work. Make sure you have all the electrical fixtures and appliances delivered to the job site before the electrician arrives.

Light fixtures The electrician hangs all light fixtures and ceiling fans. Have light bulbs available.

Switches and outlets The electrician installs all switches, outlets and covers. Also, doorbells and smoke detectors are connected.

> Occasionally cabinets cannot be ordered until drywall is installed and exact measurements can be taken. Your cabinet supplier will let you know if that is the case.

Appliances The electrician connects appliances, such as the microwave, the dishwasher and the range hood. Also, the well pump will be connected at this time.

Circuit breaker The electrician will install additional circuits at this time and connect the house wiring to those circuits.

This process will take one or more days depending on the size and complexity of the house. Your electrical system is now complete and functional.

CUSTOM STAIRS ARE FINISHED

The custom stair company will return to install custom stairs and railings.

Installation should not take longer than a day to complete.

IN-WALL CABLING IS FINISHED

Other in-wall cabling companies finish installing their systems. Security systems require monitors and keypads. Sound systems need speakers and intercom panels. Central vacuum companies need to install cover plates and deliver hoses and attachments. Probably no more than one day is needed to complete this process.

TRIM IS INSTALLED

Trim is what gives edges and corners a finished and decorative look. Trim can be very simple or quite elegant depending on your personal taste. The trim is one of the last things to be installed in a new home. Trim carpenters will do the following trimming tasks as necessary according to the building plan and contract.

Cabinets Kitchen and bathroom cabinets are secured in place and hardware is installed. Measurements for countertops should be made immediately following.

Doors Trim carpenters hang all interior doors and install hardware.

Trim The carpenters cut, fit and nail into place all the trim, including window trim, door trim, baseboard, base shoe, crown molding, chair rail and plate rail.

Closets Closet shelves and clothes poles are installed.

Installing trim is time consuming because each piece of trim must be carefully measured and cut so corners and joints fit snugly. The time needed to complete the trim process will vary according to the amount of trim to be installed.

SCHEDULE SHOWER DOORS AND MIRRORS INSTALLATION

Once you know the completion date for the interior of the house, you can schedule the shower doors and mirrors to be installed. Contact the suppliers and schedule a date following the completion of carpet installation.

■ Important questions to ask when scheduling shower doors and mirrors installation:

Do I need to confirm this date? If so, when?

Do I need to order or have any materials delivered to the building site?

When will installation be completed?

PLUMBING IS FINISHED

Now that floors are installed and walls are finished, the plumber can come back to finish the plumbing. All gas and plumbing connections are made and fixtures are installed.

Fixtures All sinks, toilets and faucets are installed.

Appliances Dishwashers and ice makers are connected. The sump pump is installed and connected if not already done.

Hose bibbs Hose bibbs, or outdoor spigots, are installed.

Plumbing connections Connections to water supply and sewer system are made at this time.

Gas connections The plumber will make all gas connections to fireplaces, ranges, water heaters and other appliances, as well as making connections to the gas service.

The plumbing will probably take a few days to finish. The plumbing for your home is now complete and functional.

SCHEDULE FINISH GRADE

Schedule the excavator to return to finish grade the yard. Make this date for as soon as possible.

■ Important questions to ask the excavator when scheduling finish grade:

Do I need to confirm this date? If so, when?

Do I need to order or have any materials delivered to the building site?

When will finish grade be completed?

SCHEDULE DRIVEWAY, SIDEWALKS AND PATIOS INSTALLATION

As soon as you know the day the finish grade will be completed, schedule a day to install the driveway, sidewalks and patios. The subcontractor you hired for this installation will depend on the types of finishes that are in your plans.

■ Important questions to ask when scheduling installation of driveway, sidewalks and patios:

Do I need to confirm this date? If so, when?

Do I need to order or have any materials delivered to the building site?

How long before I can use my driveway?

When will driveway, sidewalks and patios be completed?

SCHEDULE LANDSCAPING

As soon as you know the date the driveway will be installed, you can call the landscaper to schedule a date to begin landscaping. Make this date for after the driveway has had time to cure.

■ Important questions to ask the landscaper when scheduling landscaping:

Do I need to confirm this date? If so, when?

Do I need to order or have any materials delivered to the building site?

When will landscaping be completed?

SCHEDULE CARPETING INSTALLATION

Contact your carpet supplier and schedule a day to have your carpeting installed. Make this date for about one week in the future. This allows enough time to create a punch list and get necessary repairs made.

■ Important questions to ask the supplier when scheduling carpet installation:

Do I need to confirm this date? If so, when?

Do I need to order or have any materials delivered to the building site?

When will carpet installation be completed?

CREATE A PUNCH LIST AND SCHEDULE SUBCONTRACTORS

A punch list is a list of all necessary repairs or adjustments that need to be made. Spend quite a lot of time walking through your home inspecting the finished products.

Drywall Drywall is easily damaged, and cracks and nail pops do occur. Inspect the drywall carefully and mark the areas needing repair.

Switches and outlets Check switches and outlets to make sure they work and that they have switch plates or outlet covers.

Appliances Check to see that all of your appliances work.

Plumbing Flush all toilets and run all faucets to check for leaks.

Trim Check every piece of trim to make sure it is nailed tightly into place. Make sure the doors fit properly and all the doorknobs work. Also see that cabinet doors line up and all the accessories you ordered are there.

Exterior Check exterior finishes for completeness and especially look to see that all caulking has been done.

Special systems Check all electronic or specialty systems in your home. Try out the central vacuum, sound and security systems. Use keyless entry pads and garage door remotes.

Paint Paint and stain should be touched up after the drywall is repaired and carpet is installed.

For each repair or adjustment that needs to be made, contact the appropriate subcontractor and schedule the necessary work as soon as possible. As soon as the duties on the punch list are complete, carpet can be installed.

CARPETING IS INSTALLED

The carpet is the last major indoor building material to be installed in new home construction. The carpet supplier will install carpeting. First, a tack strip is put around the perimeter of all carpeted rooms. Next, a carpet pad is stapled down. This protects the carpet and provides extra cushioning. Then the carpet is cut to fit, stretched over the tack strip and secured. It could take several days to install carpeting depending on the number of rooms that are being carpeted.

STEP TWELVE CHECKLIST

_____ 1. Walls, ceilings and trim were painted and stained according to plan specifications. Other interior wall finishes were also installed.

_____ 2. All flooring except carpeting was installed according to plan specifications.

_____ 3. HVAC registers were installed and the thermostat was set according to my personal preference.

_____ 4. Electrical outlets, switches and light fixtures were installed according to plan specifications. Also, connections were made to appliances and the circuit box.

_____ 5. Custom stairs were installed.

_____ 6. In-wall cabling was finished according to plan specifications. Instructions and supplies for using these systems are in my possession.

_____ 7. Trim, cabinets and countertops were delivered and stocked in the designated location.

_____ 8. I have scheduled shower doors and mirrors to be installed following carpeting installation.

_____ 9. Kitchen and bathroom cabinets were set. Accurate measurements for countertops have been taken and countertops ordered for installation.

_____ 10. Doors, hardware and trim were installed. Towel bars, doorstops, closet shelves, clothes poles and other hardware were installed.

_____ 11. Plumbing fixtures were set. Connections to appliances, the water source, sewer lines and gas fixtures were made.

_____ 12. I have scheduled the excavator to finish grade the property.

— *continued* —

STEP TWELVE CHECKLIST, CONTINUED

_____ 13. I have scheduled the necessary subcontractors to install the driveway, sidewalks and patios following finish grade.

_____ 14. I have scheduled landscaping to follow driveway, sidewalks and patios installation. I have allowed ample time for the driveway to cure.

_____ 15. I have scheduled carpeting installation to follow completion of punch list repairs and adjustments.

_____ 16. I have created a punch list and contacted each subcontractor as necessary to make repairs and adjustments.

_____ 17. The punch list is satisfactorily completed.

_____ 18. The carpeting is installed according to building plan specifications.

NOTES

STEP 13
FINISHING CONSTRUCTION

- *Shower doors and mirrors are installed*
- *Final grade is finished*
- *Driveway, sidewalks and patios are installed*
- *Landscaping is finished*
- *Clean job site and house*
- *Call for inspection*
- *Request final draw*

This is the last step that you will be performing as a general contractor. Ensuring that your property is as attractive as the house is important and often necessary to receive an occupancy permit and the final draw. Do the tasks in this step in order.

SHOWER DOORS AND MIRRORS ARE INSTALLED

The shower doors and mirrors suppliers will install these items according to specifications on the building plan. These items are often forgotten by homeowners, but are important to your final product. The suppliers can complete this task in less than a day.

FINAL GRADE IS FINISHED

Your excavator will come back one last time to compete the final grade. The topsoil that was scraped from the building site and held in reserve will now be spread around the home and property. The excavator will create a gentle slope so that water runs away from the foundation. This process should not take more than a day to complete. Your property is now ready for landscaping.

DRIVEWAY, SIDEWALKS AND PATIOS ARE INSTALLED

The type of finish used for driveway, sidewalks and patios will determine the subcontractors you need and the length of time necessary to complete the work. If concrete is being used, your concrete subcontractor will form the area, pour and level the concrete, and allow it to cure before stripping the forms.

Other finishes might include blacktop, paving bricks or natural materials. The landscaper may install some of these finishes. Once the driveway, sidewalks and patios are finished, landscaping can begin.

LANDSCAPING IS FINISHED

Your geographic location will help determine the type of landscaping that must be done before an occupancy permit will be issued. Most areas will require that the yard be raked, seeded and mulched to prevent soil run-off. The time needed to complete landscaping will depend on the amount of landscaping you contracted to have done.

CLEAN JOB SITE AND HOUSE

Clean up any debris left outside the house and have the dumpster and portable toilet removed if they are still there. Clean the labels off the windows and appliances. Dust, scrub and vacuum the entire house in preparation for the final inspection and moving day.

Stamped concrete is a low maintenance alternative to paving bricks.

CALL FOR INSPECTION

The building inspector does one last check before issuing an occupancy permit. Try to be there for this inspection so that you can have anything that needs attention taken care of immediately.

The inspector will check to see that:

- all electricals are operational
- smoke detectors are operational
- all the plumbing works.
- house is complete according to plan

REQUEST FINAL DRAW

Collect all the invoices for work that has been completed and materials that have been installed. For this final draw, that includes all remaining invoices. Verify that the invoices are accurate and submit a request for a draw to your lender. Your lender will send a representative to the building site to verify that the work is complete as documented and to review your final occupancy permit. You will be notified when the funds are released. Contact the subcontractors and suppliers who submitted invoices and tell them they can receive payment when lien waivers are presented. Put copies of the lien waivers, invoices and receipts in your organizational binder.

After the final draw, it is time to roll your construction loan into a mortgage. Take care of all remaining financial details. If there is more paperwork, be sure to file copies in your organizational binder.

STEP THIRTEEN CHECKLIST

_____ 1. Shower doors and mirrors were installed.

_____ 2. The final grade was completed using reserved topsoil. The grade is a gentle slope away from the foundation.

_____ 3. Driveway, sidewalks and patios were installed according to plan specifications.

_____ 4. Landscaping was installed according to plan specifications.

_____ 5. I have completely cleaned the interior of the house.

_____ 6. I have cleaned debris from my property. All extra building materials have been disposed of or returned, and the portable toilet and dumpster have been removed.

_____ 7. The house has passed inspection and I have obtained an occupancy permit.

_____ 8. All remaining invoices have been collected and I have submitted the necessary paperwork to obtain a final draw.

_____ 9. The construction loan has been rolled into a permanent mortgage and I have filed additional financial paperwork in my organizational binder.

NOTES

STEP 14
WRAPPING IT UP

- *Contact telephone, cable TV and other utility companies*
- *Change locks and remove lockbox*
- *Set codes for keyless entry, garage remote and security system*
- *Install house numbers and mailbox*
- *Contact waste disposal service*
- *Move in*

You now return to your role as homeowner as you prepare to move into your new home. The tasks in this step can occur in any order, but you will probably want your utilities connected before moving in.

CONTACT TELEPHONE, CABLE TV AND OTHER UTILITY COMPANIES

Your house has already been wired for telephone and cable TV, but you must now contact both companies to tell them you are ready for service. They will tell you when they will make the connections. Also contact any other utilities as necessary for service.

CHANGE LOCKS AND REMOVE LOCKBOX

You had temporary lock sets installed on your exterior doors earlier. These were fine for the construction stage when damage could occur, but it is time to put the permanent lock sets on your doors. Ask your trim carpenter to change the locks and remove the lockbox.

SET CODES FOR KEYLESS ENTRY, GARAGE REMOTE AND SECURITY SYSTEM

Your garage door and security system each came programmed with a code. The subcontractors who installed these systems should have provided you with instructions to change these codes whenever you choose. Change the codes now for your protection.

INSTALL HOUSE NUMBERS AND MAILBOX

Pick house numbers that are easy to read from the road, but complement your house style. You can buy house numbers and letters from hardware stores and lumberyards. Display your address in a prominent spot on your property. Next to the front door or above a garage door are popular locations.

CONTACT WASTE DISPOSAL SERVICE

If your area has garbage and recycling services, you will need to contact them to begin trash pickup. Some municipalities have special containers that trash must be put in at the curb, and you will want to have these as soon as possible.

MOVE IN

Finally all your hard work has paid off, and the house you have always wanted is completed. You have saved yourself money, maintained decision-making control and gained personal satisfaction by general contracting your own home. Now it is time to put your feet up, relax and enjoy your new surroundings.

> Purchase a mailbox if one is needed and position it according to your post office specifications.

STEP FOURTEEN CHECKLIST

_____ 1. I have contacted the telephone, cable TV and other utility companies to have my service connected.

_____ 2. I have removed the locks and lockbox from my home and have installed the permanent lock sets.

_____ 3. I have reprogrammed my garage remotes, keyless entry and security system.

_____ 4. I have installed house numbers and a mailbox. I have given the post office and other individuals my new address.

_____ 5. I have contacted the waste disposal service for pickup.

_____ 6. I am ready to move into my new home.

NOTES

PART 3

APPENDIXES

APPENDIX A
Telephone and Address List

APPENDIX B
Cost Breakdown

APPENDIX C
Construction Calendar

APPENDIX D
Utility Hotline Numbers

INDEX

APPENDIX A

TELEPHONE AND ADDRESS LIST FOR CONTRACTORS AND SUPPLIERS

Contractor/supplier　　　　　　Address　　　　　　　　　　　　Telephone

Contractor/supplier	Address	Telephone

APPENDIX B

COST BREAKDOWN

DESCRIPTION	CONTRACTOR/SUPPLIER	LABOR	MATERIALS	TOTAL
Land				
Surveys				
House plans				
Financing fees				
Construction loan interest				
Building permits				
Utility connections				
Water supply				
Well pump				
Percolation test				
Sanitary system				
Landscaping				
Driveway, sidewalks				
Patios and decks				
Garage				
Overhead door				
Floor				
Finished interior				

DESCRIPTION	CONTRACTOR/SUPPLIER	LABOR	MATERIALS	TOTAL
Excavation				
Backfill				
Rough grade				
Final grade				
Foundation				
Concrete				
Masonry				
Damp proofing				
Insulating				
Steel beams				
Framing carpentry				
Rough lumber				
Windows and doors				
Roof				
Exterior finishes				
Masonry				
Siding				
Painting/staining				
Other _____				
Gutters				

COST BREAKDOWN

DESCRIPTION	CONTRACTOR/SUPPLIER	LABOR	MATERIALS	TOTAL
Fireplace				
Plumbing				
Electrical				
Light fixtures				
HVAC				
Insulation				
Drywall				
Interior wall finishes				
Painting/staining				
Wallpaper				
Plaster				
Other _____				
Flooring				
Hardwood				
Linoleum				
Tile				
Carpeting				
Other _____				
Trim Carpentry				

DESCRIPTION	CONTRACTOR/SUPPLIER	LABOR	MATERIALS	TOTAL
Trim lumber				
Cabinets				
Countertops				
Custom stairs				
Appliances				
Specialty systems				
Central vauum				
Sound system				
Security system				
Shower doors				
Mirrors				
Finishing touches				
Window treatments				
Furniture				
Interior decorating				
Other				
Other				
Other				
Other				
Other				

APPENDIX C

CONSTRUCTION CALENDAR

Fill in the Construction Timetable, the Payment Timetable, the Ordering and Delivery Timetable, and the Inspection Timetable using the scheduling information you received from subcontractors, lenders, suppliers, building inspectors and other professionals. Then fill in a blank calendar using this information. Remember to allow adequate time for inspections and building delays due to uncontrollable factors, such as bad weather.

Construction timetables and calendar examples are provided. This example represents only a small portion of the construction sequence and the dates given will differ from your own.

CONSTRUCTION TIMETABLE

CONSTRUCTION PHASES	LEAD TIME	NO. OF DAYS
Roofing installation	2 weeks	3 days + 2
Exterior finishes installation (siding only)	2 weeks	5 days + 2
Fireplace (none)		
Sanitary system installation (city sewer)	1 week	1 day + 2
Water supply installation (city water)	1 week	1 day + 2
Rough plumbing installation	1 week	2 days
Rough HVAC installation	1 week	2 days
Rough electrical installation	1 week	3 days

PAYMENT TIMETABLE

DRAW NUMBER	SUBMIT DATE
Draw number two — pays for all framing labor and materials	Framing completion

ORDERING AND DELIVERY TIMETABLE

SUPPLIES AND MATERIALS	ORDERING LEAD TIME	DELIVERY DATE
Dumpster	1 day	When framing begins

INSPECTION TIMETABLE

INSPECTION	GC OR SUBCONTRACTOR CONTACT
Electric service	Electrician contacts

MONTH: Construction calendar example using timetable information provided on page 156.

1	2	3	4	5	6	7
	Order dumpster	Framing begins →	Dumpster delivered	Schedule roofing and siding for 19th. Order materials delivery for 18th	Electric service installed and inspected. Contact gas and electric companies	

8	9	10	11	12	13	14
	Framing cont. →		Windows and ext. doors delivered	Schedule water, sewer and rough plumbing for 19th	Schedule rough HVAC for 20th	

15	16	17	18	19	20	21
	Locks and lockbox installed. Framing complete	Schedule electrical for 24th. Framing reserve days →	Request second draw. Roofing and siding materials delivered	City water installed. City sewer installed. Rough plumbing. Roofing begins. Siding begins	Rough HVAC begins	

22	23	24	25	26	27	28
	City water reserve days. City sewer reserve days. Roofing complete. Siding cont. →	Rough electrical begins. Roofing reserve days →	Siding reserve days	Rough electrical complete	Gas and electric connections made	

29	30	31				
		Siding reserve days				

APPENDIX C 157

CONSTRUCTION TIMETABLE

CONSTRUCTION PHASE	LEAD TIME	NO. OF DAYS
Building site preparation		
Footings installation		
Foundation walls installation		
Damp proofing and insulation		
Drain tile installation		
Steel beams installation		
Under-floor plumbing installation		
Other under-floor utilities (slab)		
Backfill and rough grade		
Foundation floor installation		
Framing		
Electric service installation		
Roofing installation		
Exterior finishes installation		
Fireplace installation		
Sanitary system installation		
Water supply installation		
Rough plumbing installation		
Rough HVAC installation		
Rough electrical installation		
Other in-wall cabling installation		
Garage floor installation		
Insulation installation		
Drywall installation		
Garage door installation		
Gutters installation		
Interior painting and staining		
Flooring installation		

CONSTRUCTION PHASE	LEAD TIME	NO. OF DAYS
Finish HVAC.	_____	_____
Finish electrical.	_____	_____
Custom stairs installation.	_____	_____
Finish other in-wall cabling	_____	_____
Trim installation.	_____	_____
Finish plumbing.	_____	_____
Punch list completed	_____	_____
Carpeting installation.	_____	_____
Shower doors installation.	_____	_____
Mirrors installation.	_____	_____
Final grade	_____	_____
Driveway installation.	_____	_____
Sidewalks and patios installation	_____	_____
Landscaping	_____	_____
Other: _____	_____	_____
Other: _____	_____	_____
Other: _____	_____	_____
Other: _____	_____	_____
Other: _____	_____	_____
Other: _____	_____	_____

PAYMENT TIMETABLE

DRAW NUMBER	SUBMIT DATE
Draw One.	_____
Draw Two.	_____
Draw Three.	_____
Draw Four	_____
Draw Five.	_____
Draw Six.	_____

ORDERING AND DELIVERY TIMETABLE

SUPPLIES AND MATERIALS	ORDERING LEAD TIME	DELIVERY DATE
Trusses	_____	_____
Windows and exterior doors	_____	_____
Framing lumber	_____	_____
Portable toilet	_____	_____
Steel beams	_____	_____
Dumpster	_____	_____
Cabinets and countertops	_____	_____
Interior trim	_____	_____
Light fixtures	_____	_____
Appliances	_____	_____
Other: _____	_____	_____
Other: _____	_____	_____
Other: _____	_____	_____

INSPECTION TIMETABLE

INSPECTION	GC OR SUBCONTRACTOR CONTACT
Footings	_____
Foundation	_____
Under foundation floor utilities	_____
Foundation floor	_____
Electric service	_____
Framing and rough mechanicals	_____
Insulation	_____
Final	_____
Other: _____	_____
Other: _____	_____
Other: _____	_____
Other: _____	_____

MONTH:

NOTE: Buy a blank calendar or make copies of this page as necessary.

APPENDIX C

APPENDIX D

UTILITY HOTLINE NUMBERS
Source: American Public Works Association

STATE	CENTER	PHONE
Alabama	Alabama Line Location Center	800-292-8525
Alaska	Locate Call Center of Alaska	800-478-3121
Arizona	Arizona Blue Stake	800-782-5348
Arkansas	Arkansas One-Call System	800-482-8998
California	Underground Service Alert	800-227-2600
Colorado	Utility Notification Center of Colorado	800-922-1987
Connecticut	Call Before You Dig	800-922-4455
Delaware	Miss Utility of Delaware	800-282-8555
Florida	Call Sunshine	800-432-4770
Georgia	Utilities Protection Center	800-282-7411
Hawaii	Underground Service Alert	800-227-2600
Idaho	Palouse Empire Underground	800-822-1974
	Utilities Underground Location Center	800-424-5555
	Dig Line	800-342-1585
	One-Call Concepts – Idaho	800-626-4950
	Shoshone County One-Call	800-398-3285
	PASS WORD	800-428-4950
Illinois	JULIE	800-892-0123
	Digger (Chicago Utility Alert Network)	312-744-7000
Indiana	Indiana Underground Plant Protection	800-382-5544
Iowa	Underground Plant Location Service	800-292-8989
Kansas	Kansas One-Call Center	800-DIG-SAFE
Kentucky	Kentucky Underground Protection	800-752-6007
Louisiana	Louisiana One-Call System	800-272-3020
Maine	Dig Safe – Maine	888-DIG-SAFE
Maryland	Miss Utility	800-257-7777
	Miss Utility of DELMARVA	800-282-8555
Massachusetts	Dig Safe – Massachusetts	888-DIG-SAFE
Michigan	Miss Dig System	800-482-7171
Minnesota	Gopher State One-Call	800-252-1166
Mississippi	Mississippi One-Call System	800-227-6477
Missouri	Missouri One-Call System	800-344-7483
Montana	Utilities Underground Location Center	800-424-5555
	Montana One-Call Center	800-551-8344

STATE	CENTER	PHONE
Nebraska	Diggers Hotline	800-331-5666
Nevada	Underground Service Alert North	800-227-2600
New Hampshire	Dig Safe – New Hampshire	888-DIG-SAFE
New Jersey	Garden State Underground	800-272-1000
New Mexico	New Mexico One-Call System	800-321-ALERT
	Las Cruces-Dona Ana Utility Council	800-526-0400
New York	Underground Facilities Protective	800-962-7962
	New York City-Long Island One-Call	800-272-4480
North Carolina	North Carolina One-Call Center	800-632-4949
North Dakota	Utilities Underground Location Center	800-795-0555
Ohio	Ohio Utilities Protection Service	800-362-2764
	Oil and Gas Producers Underground Protection	800-925-0988
Oklahoma	Call Okie	800-522-6543
Oregon	Oregon Utilities Notification Center	800-332-2344
Pennsylvania	Pennsylvania One-Call System	800-242-1776
Rhode Island	Dig Safe – Rhode Island	888-DIG-SAFE
South Carolina	Palmetto Utility Protection Service	888-721-7877
South Dakota	South Dakota One-Call	800 781-7474
Tennessee	Tennessee One-Call System	800-351-1111
Texas	Texas One-Call System	800-245-4545
	Texas Excavation Safety System	800-344-8377
	Lone Star Notification System	800-669-8344
Utah	Blue Stakes Location Center	800-662-4111
Vermont	Dig Safe – Vermont	888-DIG-SAFE
Virginia	Miss Utility of Virginia	800-552-7001
	Miss Utility	800-257-7777
	Miss Utility of DELMARVA	800-441-8355
Washington	Underground Utilities Notification Center	800-424-5555
	Grays Harbor & Pacific County Utility	206-532-3550
	Utilities Council of Cowlitz County	360-425-2506
	Chelan-Douglas Utilities Coordinating	509-663-6111
	Upper Yakima County Underground	800-553-4344
	Inland Empire Utility Coordinating	509-456-8000
	Utilities Notification Center	800-332-2344
West Virginia	Miss Utility of West Virginia	800-245-4848
Wisconsin	Diggers Hotline	800-242-8511
Wyoming	Wyoming One-Call	800-348-1030
	Wyoming Association of Local Utility Coordinating Councils	800-849-2476
D. C.	Miss Utility	800-257-7777

INDEX

A

Air conditioning. *See* HVAC
Anchoring system 82, 94
Appliances 51, 122, 129, 130, 131
Approval, permits. *See* Building permits
Architect 28
Attorney.......................... 27, 37

B

Backfill........................... 58, 85
Basement 80-89
Better Business Bureau................. 48
Bids 41-47, 49-52
Blueprints. *See* House plans
Bracing 85
Brick ledge.......................... 82
Building checklists. *See* Checklists
Building contracts. *See* Contracts
Building inspections 36, 58-59
   electric service 95
   occupancy 139
   footings....................... 80-81
   foundation 83-84
   foundation floor.................... 86
   framing......................... 110
   insulation...................... 120
   rough mechanical.................. 110
   under-floor 84-85
Building inspector................ 36, 58-59
Building permits 36, 47, 52, 58-59
Building supplies................. 49-52, 69

C

Cabinets...................... 50, 118, 129
Carpentry..................... 42, 43, 131
   framing 69-70, 94
   inspection..................... 59, 110
   trimming 121, 129
Carpeting 50, 131, 132,
                 see also Flooring
Cellar. *See* Basement
Central vacuum system... 51, 99, 109, 129, 131
Chamber of Commerce 48
Change orders 6, 47, 48
Changes, construction 29, 47, 58
Checklists
   Part one 7, 22, 37, 53-54, 61
   Part two 71, 75, 88-89, 100-101, 112-113,
              123-124, 133-134, 140, 145
Checks. *See* Draws

Claims. *See* Liens
Cleaning, job site 49, 74, 89, 138
Closing, loan. *See* Financing
Construction
   calendar 59-60, 156-161
   loan. *See* Financing
   sequence............................ 5
   timetable 59, 156-161
Contingencies
   budget..................... 20, 40, 47
   property purchase.................... 27
Contracts 47-52
Copyright 29
Cost
   breakdown.................... 152-155
   estimation 19-21
Countertops.................. 50, 118, 129
Covenants 27, 29
Crawlspace 80-89
Culvert............................. 74
Curb cutting permit.................... 36

D

Damp proofing 83
Decks, or porches..................... 107
Designers, house 28
Doors............................ 70, 95
Drain tile 80, 83, 86
Draws 40-41
   fifth 139
   first............................ 87
   fourth.......................... 122
   second 99
   third........................... 111
Driveway 130, 138
   construction....................... 74
Drywall............. 42, 106, 110, 120, 131
Dumpster............................ 95

E

Electric............................. 42
   circuits...................... 95, 129
   company........................ 59, 96
   finish installation 120-121, 128-129
   in masonry walls 96
   inspections..................... 59, 95
   permit 36
   rough installation 98, 109
   service......................... 94, 95
   under-floor..................... 81, 84
Elevation 28, 68

Estimates. *See* Cost, estimation
Excavation 41-42, 73-75
   finish grade. 130, 138
   inspection. 58
   rough grade 85
   site preparation 68
Exterior doors. *See* Doors
Exterior finishes. 42, 131
   fascia 107
   installation 106-107
   masonry 96, 107
   scheduling. 96-97
   siding 107
   soffit. 107
   stucco. 107

F

Fascia. *See* Exterior finishes
Financing, ... 40-41, 52, 139
   closing loan. 18, 27, 52, 139
   construction loan 18, 40
   draws. *See* Draws
   mortgage 16-19
   pre-qualify 16-19
Finish grade. *See* Excavation
Fireplace 51, 80, 97, 98, 107, 108
Flooring 50, 118, 128, 131, 132
Footings. *See* Foundation
Foundation 42, 78-89
   damp proofing 83
   floor 86, 87
   footings 74, 80-81
   height. 68
   inspection 58-59, 83-84, 86
   insulation 83, 86
   preparation. 74
   scheduling. 68-69
   utility lines through 69, 82, 84
   walls. 82
Framing. *See* Carpentry
Furnace. *See* HVAC

G

Garage
   door 111, 121, 144
   floor 110, 111, 119
Gas company. 59, 96
                              See also Plumbing
General contractor 4-7
Grade. *See* Excavation
Gutters. 111, 121

H

Heating. *See* HVAC
House plans 27

blueprints 28, 29, 40, 41, 49
evaluating. 29
locating 28-29
worksheets. 10-14
HVAC
   ductwork 108
   finish installation. 120-121, 128
   inspection. 59, 110
   permit 36
   rough installation 98, 108-109
   under-floor 81, 84
   vents. 109

I

Inspections. *See* Building, inspections
Insulation 42, 110, 119-120
   foundation floor. 86
   inspection. 120
Insurance
   homeowners. 52
   public liability. 46, 48, 52
   worker's compensation 46, 48
Interior wall finishes. 44, 118, 128
   plaster. 120
   wallpaper 118, 128

L

Land. *See* Property
Land features worksheet 14-16
Landscaping. 131, 138
Lender payment schedule. *See* Draws
Lien waiver 18-19
Liens. 18-19
Lifestyle worksheet 10-12
Light fixtures 50, 122, 128
Loans. *See* Financing
Lockbox 95, 99, 144
Locks 95, 99, 144
Lot. *See* Property
Lumber 50, 69-70

M

Masonry. 96
         See also Foundations, Exterior finishes
Mirrors 51, 129-130, 138
Mortgage. *See* Financing

N

New home features worksheet 12-14

O

Occupancy permit. *See* Building permits
OSHA. 49

P

Painting 44, 107, 118, 128, 131,
   See also Exterior finishes, Interior wall finishes
Patios 130, 138
Percolation test 27
Permits. *See* Building permits
Plans. *See* House plans
Plaster. *See* Interior wall finishes
Plat. *See* Surveys
Plot plan. *See* Surveys
Plumbing 42, 131
   finish installation 121, 130
   gas 108, 130
   in masonry walls 96
   inspection 59, 110
   permit 36
   rough installation 98, 108
   under-floor 81-82, 84-85
Porches, or decks 107
Pre-qualify. *See* Financing
Property 29
   contingencies 27
   evaluation 27
   locating 26-27
   purchasing 52
Proposals. *See* Contracts
Punch list 131-132

R

Real Estate. *See* Property
Rebar 80, 81
Record keeping 3
References 46, 48
Roof 42
   framing 94-95
   installation 96, 106
   vents 106
Rough grade. *See* Excavation

S

Safety 49, 74
Sanitary system 27, 42, 97, 107
   inspections 59, 130
   permit 36
                     See also Plumbing
Security system 51, 99, 109, 129, 131, 144
Septic system. *See* Sanitary system
Sewer. *See* Sanitary system
Shower doors 51, 129, 130, 138
Sidewalks 130, 138
Siding. *See* Exterior finishes
Site preparation. *See* Excavation
Slab foundation 80-89
Soffit. *See* Exterior finishes

Soil erosion permit 36
Sound system 51, 99, 109, 129, 131
Specifications 29-35
Staining. *See* Painting
Stairs 50, 95, 96, 120-121, 129
Steel beams 50, 82-83, 86-87
Stock plans. *See* House plans
Subcontractors 41-46
Surveys 52, 74, 87

T

Telephone 59, 96, 98, 109, 144
Television 59, 96, 98, 109, 144
Termites 85
Timetables. *See* Construction timetables
Toilet, portable 70
Topsoil 68, 74, 138
Trenches 74
Trim 70, 118, 129
Trusses 50, 70, 94
TV. *See* Television

U

Utilities 59, 68, 70, 109, 144, 162-163

V

Vapor barrier 86, 119
Ventilation. *See* HVAC
Vinyl. *See* Flooring

W

Wallpaper. *See* Interior wall finishes
Warranties 46, 48, 51
Water supply system 27, 42, 97, 108, 130
   inspection 59, 108
   permit 36
                     See also Plumbing
Well. *See* Water supply system
Windows 50, 70, 95
Worksheets
   construction calendar 156-161
   cost breakdown 152-155
   cost estimation 20-21
   land features 14-16
   lifestyle 10-12
   new home features 12-14
   pre-qualify for financing 16-17
   pre-qualifying lenders 17-18
   specification 30-35
   subcontractors . . . . ,44-46

Z

Zoning 27, 36, 74
   permit 36

ORDER FORM

Postal Orders . . . . . . Baine Books
PO Box 892
East Troy, WI 53120

Fax Orders . . . . (414) 642-5975

Telephone Orders . . . (877) 642-3390 (Toll-free)

Online Orders . . . . . . www.BaineBooks.com

SHIP TO

Name _____

Address _____

City _____ State _____ Zip _____

Telephone (_____) _____

QUANTITY

CONTRACTING DETAILS:
*A do-it-yourself construction schedule
and homebuilding handbook.* . . . . _____ at **19.95 each** . . . _____

    Shipping: post-paid within the US. . . . . __**FREE**__

    WI residents add 5.5% sales tax . . . . _____

    TOTAL OF ORDER . . . . _____

PAYMENT

☐ Check ☐ Money order

☐ Credit Card . . . . ☐ VISA . . . . ☐ MasterCard

Card number _____

Name on card _____ Exp. date _____

Signature _____

THANK YOU!